Cultural Heritage and Tourism in the D

Cultural Heritage and Tourism in the Developing World is the first book of its kind to synthesize global and regional issues, challenges, and practices related to cultural heritage and tourism, specifically in less-developed nations. The importance of preservation and management of cultural heritage has been realized as an increasing number of tourists are visiting heritage attractions. Although many of the issues and challenges developing countries face in terms of heritage management are quite different from those in the developed world, there is a lack of consolidated research on this important subject. This seminal book tackles the issues through theoretical discourse, ideas, and problems that underlay heritage tourism in terms of conservation, management, economics and underdevelopment, politics and power, resource utilization, colonialism, and various other antecedent notions that have shaped the development of heritage tourism in the less-developed regions of the world.

The book is composed of two sections. The first section highlights the broader conceptual underpinnings, debates, and paradigms in the realm of heritage tourism in developing regions. The chapters in this section examine heritage resources and the tourism product; protecting heritage relics, places and traditions; politics of heritage; and the impacts of heritage tourism. The second section examines heritage tourism issues in specific regions, including the Pacific Islands, South Asia, the Caribbean, China and Northeast Asia, Southeast Asia, subSaharan Africa, Central and Eastern Europe, the Middle East and North Africa, and Latin America. Each region has unique histories, cultures, political traditions, heritages, issues, and problems, and the way these issues are tackled vary from place to place.

This volume develops frameworks that are useful tools for heritage managers, planners and policy-makers, researchers, and students in understanding the complexity of cultural heritage and tourism in the developing world. Unlike many other books written about developing regions, this book provides insiders' perspectives, as most of the empirical chapters are authored by the individuals who live or have lived in the various regions and have a greater understanding of the region's culture, history, and operational frameworks in the realm of cultural heritage. The richness of this "indigenous" or expert knowledge comes through as each regional overview elucidates the primary challenges and opportunities facing heritage and tourism managers in the less affluent areas of the world.

Dallen J. Timothy is Professor of Community Resources and Development at Arizona State University, USA. He is Editor of the Journal of Heritage Tourism and also researches tourism issues related to religion, developing countries, planning, and borders.

Gyan P. Nyaupane is the graduate program director and assistant professor at Arizona State University, USA. He has research interests in heritage management, conservation, and tourism development in the developing world

Contemporary geographies of leisure, tourism and mobility

Series editor: C. Michael Hall

Professor, Department of Management, College of Business & Economics, University of Canterbury, Christchurch, New Zealand

The aim of this series is to explore and communicate the intersections and relationships between leisure, tourism, and human mobility within the social sciences.

It will incorporate both traditional and new perspectives on leisure and tourism from contemporary geography, e.g., notions of identity, representation and culture, while also providing for perspectives from cognate areas such as anthropology, cultural studies, gastronomy and food studies, marketing, policy studies and political economy, regional and urban planning, and sociology, within the development of an integrated field of leisure and tourism studies.

Also, increasingly, tourism and leisure are regarded as steps in a continuum of human mobility. Inclusion of mobility in the series offers the prospect to examine the relationship between tourism and migration, the sojourner, educational travel, and second home and retirement travel phenomena.

The series comprises two strands:

Contemporary geographies of leisure, tourism and mobility aims to address the needs of students and academics, and the titles will be published in hardback and paperback. Titles include:

Routledge studies in contemporary geographies of leisure, tourism and mobility is a forum for innovative new research intended for research students and academics, and the titles will be available in hardback only. Titles include:

Cultural Heritage and Tourism in the Developing World

A regional perspective

Edited by
Dallen J. Timothy and Gyan P. Nyaupane

Routledge
Taylor & Francis Group

LONDON AND NEW YORK

First published 2009
by Routledge
2 Park Square, Milton Park, Abingdon, Oxon OX14 4RN

Simultaneously published in the USA and Canada
by Routledge
270 Madison Avenue, New York, NY 10016

Routledge is an imprint of the Taylor & Francis Group, an informa business

Typeset in Times New Roman by
Taylor & Francis Books
Printed and bound in Great Britain by
CPI Antony Rowe, Chippenham, Wiltshire

British Library Cataloguing in Publication Data
A catalogue record for this book is available from the British Library

Library of Congress Cataloging in Publication Data
Cultural heritage and tourism in the developing world: a regional
perspective / Dallen J. Timothy and Gyan P. Nyaupane (eds)
    p. cm. — (Contemporary geographies of leisure, tourism and mobilities)
Includes bibliographical references and index.
[etc.]
    1. Heritage tourism—Developing countries. I. Timothy, Dallen J. II.
Nyaupane, Gyan P., 1968–
    G156.5.H47C853 2009
    338.4'791091724—dc22
                            2008048520

ISBN10: 0-415-77621-X (hbk)
ISBN10: 0-415-77622-8 (pbk)
ISBN10: 0-203-87775-6 (ebk)

ISBN13: 978-0-415-77621-9 (hbk)
ISBN13: 978-0-415-77622-6 (pbk)
ISBN13: 978-0-203-87775-3 (ebk)

Contents

Figures

Tables

Contributors

Megha Budruk is Assistant Professor of Community Resources and Development at Arizona State University, Phoenix, USA.

Mariusz Czepczyński is Senior Lecturer in the Department of Economic Geography, University of Gdansk, Gdynia, Poland.

Rami F. Daher is Associate Professor of Architecture at German-Jordanian University, Amman, Jordan, and Principal of TURATH: Architecture and Urban Design Consultants, Amman, Jordan.

David T. Duval is Associate Professor in the Department of Tourism at the University of Otago, Dunedin, New Zealand.

C. Michael Hall is Professor of Management at the University of Canterbury, Christchurch, New Zealand.

Joan C. Henderson is an Associate Professor in the Nanyang Business School, Singapore.

Leslie-Ann Jordan is a Lecturer in Hospitality and Tourism at the University of the West Indies, Trinidad.

Duncan Light is Associate Professor of Human Geography, Liverpool Hope University, Liverpool, UK.

Oyunchimeg Luvsandavaajav is Lecturer in Human Geography and Tourism at the National University of Mongolia, Ulaanbaatar, Mongolia.

Gyan P. Nyaupane is Assistant Professor of Community Resources and Development at Arizona State University, Phoenix, USA.

Regina Schlüter is the Director of Centro de Investigaciones y Estudios Turísticos, Buenos Aires, Argentina.

Victor B. Teye is Associate Professor of Community Resources and Development at Arizona State University, Phoenix, USA.

Dallen J. Timothy is Professor of Community Resources and Development at Arizona State University, Phoenix, USA.

Bihu Wu is Professor in the Department of Urban and Regional Planning, Peking University, Beijing, China.

Craig Young lectures at the Manchester Institute of Social and Spatial Transformations, Manchester Metropolitan University, UK.

Acknowledgments

The editors would like to extend a warm thank you to Andrew Mould and Michael P. Jones at Taylor and Francis for their patience and perseverance. They have been a pleasure to work with, and their enthusiasm for the project was very encouraging. We would also like to thank the contributors for their expertise in putting together a set of well-written chapters on regional issues in heritage tourism. Their efforts are much appreciated, especially those who stepped in at the last minute to write chapters that were initially meant to be written by others. The efforts of Bharath Sollapuram and Surya Poudel in drawing the maps are gratefully acknowledged. Finally, our biggest debt of gratitude, as always, goes to our wives (Carol and Meera) and children who have put up with late nights and shouldered so much more than their fair share of work during the preparation of this book.

Preface

This book emerged from a long-held belief by the editors and many other academic observers that many tourism dynamics in the developing world are quite different from those in the more affluent portions of the globe. The editors have lived in, traveled around, and worked widely throughout the developing world and have observed first hand many of these unique dimensions of tourism. Both editors have a strong interest in cultural heritage and have studied heritage issues in the developing world context on many occasions. Based on an understanding of the different heritage and tourism experiences in the less-developed world, this book was conceived. Our original plan was to provide conceptual overviews of critical issues, challenges, and opportunities in the heritage tourism context in developing regions and then invite scholars who themselves are interested in heritage tourism and who live in less-affluent countries of the world to contribute regional overviews to highlight these crucial issues in their various regions. This "indigenous" approach was considered the best way to understand the challenges and issues facing heritage places in developing countries. Unfortunately, however, several people were unable to contribute for various reasons, and new authors were asked to fill in, some in the last hours of the project. We are grateful to them and their willingness to work with us. Despite this glitch, we were able to put together a successful collection of regional overviews that illustrate the common problems and prospects of cultural heritage and tourism in the less-affluent regions. The developing world is rich in awesome cultural resources that have created and continue to create some of the most attractive tourist destinations in the world.

Dallen J. Timothy
Gilbert, Arizona, USA

Gyan P. Nyaupane
Tempe, Arizona, USA

Part I

Heritage issues and challenges in developing regions

1 Introduction

Heritage tourism and the less-developed world

People visiting cultural and historical resources is one of the largest, most pervasive, and fastest growing sectors of the tourism industry today. In fact, heritage tourism appears to be growing much faster than all other forms of tourism, particularly in the developing world, and is thus viewed as an important potential tool for poverty alleviation and community economic development (UNWTO 2005). Heritage tourism typically relies on living and built elements of culture and refers to the use of the tangible and intangible past as a tourism resource. It encompasses existing cultures and folkways of today, for they too are inheritances from the past; other immaterial heritage elements, such as music, dance, language, religion, foodways and cuisine, artistic traditions, and festivals; and material vestiges of the built cultural environment, including monuments, historic public buildings and homes, farms, castles and cathedrals, museums, and archeological ruins and relics. Although the heritage industry has in the past focused overwhelmingly on the patrimony of the privileged (e.g., castles, cathedrals, stately homes), there is now widespread acknowledgment and acceptance of everyday landscapes that depict the lives of ordinary people: families, farmers, factory workers, miners, fishers, women and children (Timothy and Boyd 2006a).

There is recognition in tourism studies in general, and heritage tourism in particular, that tourism and its impacts, constraints, and management implications are different in the developing world from conditions in the developed world. These differences are underscored principally by differences in economics; politics, power and empowerment; colonialism; conservation/preservation practices; social mores; cultural vitality; gender and socio-economic disparities; urbanization; and legislative engagement, among others (Britton 1982; Harrison 1992; Huybers 2007; Mowforth and Munt 1998; Oppermann and Chon 1997; Timothy 1999). These differences are especially perceptible in the realm of heritage tourism and its impacts (Berger 1996; Bruner 1996; Evans 1998; Leung 2001; Timothy and Boyd 2003, 2006a; Wager 1995).

In spite of the growing interest in this area, the body of knowledge is young, and there still remains a dearth of consolidated research on the dynamics of cultural heritage tourism in the developing regions of the world. This knowledge is vital for the preservation of heritage and the longevity of tourism in those destinations. This book aims to address the issues that are unique to the developing world, as well as the matters that overlap with the more developed parts of the globe. It aims to discuss much of what is known about heritage tourism in the less-developed world and to examine and challenge the existing paradigms, concepts, and practices related to cultural heritage tourism. It provides a global overview of the most critical issues facing heritage managers and heritage destinations in less-developed countries, including opportunities and prospects for developing heritage-based tourism.

Developing countries and heritage tourism

From a socio-economic perspective, the world has been, and continues to be, divided into developed and developing countries, sometimes referred to as the "haves" and "have-nots," the "North" and "South" (because of the high concentration of poorer countries in the southern hemisphere), "industrialized" versus "non-industrialized," or "more-developed" (MDCs) and "less-developed" (LDCs) states of the world. Such designations are fraught with definitional problems, as the earth continues to be a dynamic place, and many less-developed countries continue to progress on paths of development and modernization. Some nations hardly fit within the basic framework of developed and developing countries, such as a few in Eastern Europe, East Asia, the Middle East, and South America. These, according to Hobbs (2009), might best be considered "newly industrializing countries," although they all share fairly common socio-economic characteristics that define their level of development.

While there is no absolute or universally accepted set of criteria to determine where a country lies on the spectrum of development, the world community and development agencies have identified several variables that permit the distinction between the more-developed and less-developed portions of the world. Per capita gross domestic product (GDP) or per capita gross national income (GNI) are among the most important indicators. Another related indicator that is more revealing in actual terms is annual per capita gross national income purchasing power parity (GNI PPP), which takes into account GNI (GDP plus money from abroad) and differences between countries in the relative prices of goods and services (Hobbs 2009). The disparities between wealthy and poor countries are quite remarkable. According to the Population Reference Bureau (2008), the average GNI PPP in MDCs in 2007 was US$31,200, while in LDCs, the average was US$4,760. The wealthiest country, according to this index, is Luxembourg, where the per capita annual GNI PPP was approximately US$64,400. The poorest two countries, according to this per capita index, are Liberia and the Democratic Republic of the Congo, whose GNI PPP were both measured at US$290 in 2007.

The human development index (HDI) is another indicator that measures the average achievement of countries in three basic dimensions of human development, including a long and healthy life, access to knowledge, and a decent standard of living (UNDP 2008). These are measured by life expectancy at birth, adult literacy and combined gross enrollment in education, and per capita GDP. It is argued that the concept of the HDI is much broader than simple GDP. This criterion divides countries into three categories, high, medium, and low human development, and helps compare and monitor long-term trends in human development. According to the 2008 Human Development Index report (UNDP 2008), Iceland (0.968), Norway (0.968) Australia (0.961), and Canada (0.961) rank the highest, and Sierra Leone (0.336), Burkina Faso (0.370), Guinea-Bissau (0.374), and Niger (0.374) rank the lowest.

Other development indicators include, but are not limited to, birth rate, infant mortality, life expectancy, literacy rates, urban versus rural populations, levels of energy use, industrial versus service economies (de Blij and Muller 2006). Table 1.1 shows several of these indicators and their characteristics in relation to the level of development.

Many historical, socio-economic, geographical, and political factors come into play in determining the level and rate of development of any given country. Climate has long been seen as a determiner of human behavior and capability, with extreme climatic conditions being disadvantageous to growing nutritious crops in adequate abundance to support a population (de Blij and Muller 2006; Semenov and Porter 1995). Historically, people in tropical areas have tended to be less productive and poorer than people living in colder climates because of their vulnerability to heat and diseases, which is true even in the twenty-first century as the South is less affluent than the North (Landes 1998). Natural resource advantages and their distribution are also often cited as reasons why places develop or remain in an underdeveloped state (Pearce

Table 1.1 Characteristics of developed and developing countries

Traits	Developed countries (MDCs)	Developing countries (LDCs)
Per capita GDP and income	High	Low
Percentage of population employed in manufacturing	High	Low
Energy use	High	Low
Percentage of population living in cities	High	Low
Percentage of population living in rural settings	Low	High
Birth rate	Low	High
Death rate	Low	Higher than in MDCs
Population growth rate	Low	High
Percentage of population under age 15	Low	High
Percentage of population that is literate	High	Low
Amount of leisure time available	High	Low
Life expectancy	High	Low

GDP, gross domestic product.
Source: After Hobbs (2009: 44).

and Turner 1990; Sachs and Warner 1995). Accessibility and location are primary issues as well, particularly in relation to trade in natural resources. Countries with inaccessible physical geography or those that are landlocked have a tendency to lag behind countries with ocean access, deep water ports, and more extensive coastal and agricultural plains (Faye *et al.* 2004; MacKellar *et al.* 2000). According to one prominent line of thinking, European colonization and the over-exploitation of natural resources that accompanied it have resulted in a modern dependency of the colonized world on the Western, colonial powers for income, trade, and governance. This has resulted in a legacy of continued dependency relationships (neo-colonialism) between the developed and developing portions of the world (Bertocchi and Canova 2002; Crosby 2004).

From a tourism perspective, less-developed countries are extremely important as destinations and players in the global industry. Travel to and within the developing world is growing at a rapid rate, more quickly in fact than in more developed regions. Between 1990 and 2005, for example, international arrivals in developing countries grew by an average of 6.5 percent each year. Arrivals in the developed world during the same period averaged below 3 percent per annum. This remarkable growth is a result of many factors, including improved standards of living among the traveling public, increased freedom to travel within many parts of the less-developed world, improved international relations, new markets opening up (e.g., People's Republic of China and Eastern Europe), and higher priorities being placed on travel in terms of visitor and government spending.

As already noted, a salient part of this increase in travel demand is related directly to cultural heritage as a resource for tourism. Approximately 60 per cent of all of UNESCO's World Heritage Sites (natural and cultural) are located in developing nations, depending on precisely how these are defined. Some of the most spectacular remnants of ancient civilizations and contemporary colonial patrimony are located in developing regions. The Pyramids of Giza and Valley of the Kings (Egypt), Angkor Wat (Cambodia), Borobudur and Prambanan (Indonesia), the ancient city of Timbuktu (Mali), the Roman ruins of Palmyra (Syria), Great Zimbabwe Monument (Zimbabwe), Tikal (Guatemala), Lumbini, the birthplace of Buddha (Nepal), historic Istanbul (Turkey), and Dracula's Castle (Romania) are all examples of well-known and highly visible heritage places in the less-developed world.

The New Seven Wonders of the World project was initiated in 2001 by a non-government body of volunteers to determine the modern world's most spectacular cultural wonders to correspond to those of the ancient world. The organization, New7Wonders, called for a global referendum to determine the seven wonders of the modern world. This hyped-up and highly visible campaign resulted in the 2007 announcement (after apparently more than 100 million votes) of seven new wonders of the world, six of which are located in less-developed countries: Chichen Itza (Mexico), Christ Redeemer (Brazil), Great Wall of China (China), Taj Mahal (India), Petra (Jordan), and Macchu Pichu (Peru). The seventh is the Colosseum in Italy. This exercise in "global democracy" was considered by the organization to be a significant success (New7Wonders 2008) and, for the purposes of this chapter, it illustrates the importance of the developing world as a host for some of the earth's most wondrous and scientifically important historic relics.

In the not so distant past, many tropical developing nations concentrated their promotional efforts on boosting their tourism economies via the sun, sea and sand (SSS) model of unplanned and poorly regulated tourism growth. Today, however, given the traditional socio-cultural and ecological pitfalls associated with mass tourism based on the three Ss, together with the realization of the importance of cultural heritage as a resource for tourism, many traditional beach destinations have started to refocus their promotional and planning efforts to include heritage attractions to broaden their resource base and tourism offerings (Bennet 1993; Luxner 1999; McCabe 1992).

While beaches and warm climates are still the primary tourist draw, culture and heritage are becoming more important in the product mix, particularly in countries in Africa, Latin America, the South Pacific, and the Caribbean. Even the sand- and sea-dependent Turks and Caicos Islands are considering broadening their appeal to emphasize more local culture and local history (Cameron and Gatewood 2008). Irandu (2004) noted a similar situation in Kenya, where tourism has been almost exclusively centered on wildlife and beaches but is now being extended to colonial and indigenous African heritage. Many island states in the Caribbean, as well as LDCs elsewhere, have been somewhat wary about developing their colonial heritage—slavery and colonial architecture—because colonial times are rarely remembered with any degree of fondness. For many Caribbean people, the memory of slavery is still fresh, and the European rulers depleting the islands' resources then leaving the slave descendents to fare for themselves after independence is still rightfully a point of contention for many. With the Caribbean indigenous culture having all but disappeared from the islands, with the exception of a small remaining population on the island of Dominica (Slinger 2000), the primary focus of heritage is the colonial past (Bennet 1993), and this is, according to some observers (e.g., McCabe 1992), part of the reason why the Caribbean states have been reluctant to develop heritage tourism.

The heritage tourism product

Researchers and managers have a tendency to divide tourism into types as a way of facilitating research and creating knowledge, marketing, planning, and managing impacts. Just as tourism is often subdivided into forms such as nature-based, sport, heritage, health, and adventure, heritage tourism can also be divided into parts or subtypes as a way of presenting its complexities and understanding its distinctive characteristics. This is usually done from both supply and demand perspectives, in that types of heritage tourism (and tourism in general) are defined by the places, events, and artifacts observed or visited (i.e., consumed), as well as by the motives of and activities undertaken by the tourists who consume them. Research suggests that, in most cases, people visit heritage places to enhance learning, satisfy curiosity and feelings of nostalgia, grow spiritually, relax, get away from home, spend time with loved ones, or "discover themselves" (Confer and Kerstetter 2000; Krakover and Cohen 2001; Poria *et al.* 2004; Prentice *et al.* 1997; Timothy 1997; Timothy and Boyd 2003). One recent study (Nyaupane *et al.* 2006) classified heritage tourists into three types based upon their motivations: culture-focused, culture-attentive, and culture-appreciative. These motivations, combined with relics from the past, create a range of heritage tourism types that are examined in the paragraphs that follow, all of which are important constituents of the heritage product in the less-developed world.

Religious tourism is one of the most prevalent forms of heritage tourism in the developing world today and is among the earliest precursors of modern-day tourism. Pilgrimage takes many forms, but central among these is the desire of religious adherents to supplicate deity for blessings, become closer to God, offer more sincere prayers, become healed, and receive forgiveness for sins. Much pilgrimage requires self-humbling and penitence, which can be effected more readily in some cases by the afflictions associated with traveling along a prescribed pilgrim route (Shair and Karan 1979). In some religious traditions, the pathway to the religious site can be as enlightening and spiritually moving as arriving at the holy site itself (Bhardwaj 1983; Cousineau 2000; González and Medina 2003). This form of travel is required or encouraged in Islam, Hinduism, Buddhism, and Christianity, and many of the most sacred places on earth are located in the developing countries of the Middle East, South Asia, and Southeast Asia. In India, for example, domestic and international travel by Hindus for religious purposes is an important part of the tourism economy, and the Kumba Mela religious pilgrimage is the largest tourist gathering in the world (Singh 2006).

Pilgrimage should be considered a form of heritage tourism from at least three perspectives. First, the sites visited are heritage places, including churches, mosques, temples, synagogues, shrines, sacred mountains, and caves/grottos. Second, pilgrimage routes have become heritage resources based on their historical role in the practice of pilgrimage. Finally, the forms of worship and the religious rites undertaken at venerated places have become part of an intangible heritage, or a set of socio-cultural practices that demonstrate inwardly and outwardly the weightiness of the journey.

Diaspora tourism is a form of ethnic and personal heritage tourism, wherein people from various backgrounds travel to their homelands in search of their roots, to celebrate religious or ethnic festivals, to visit distant or near relatives, or to learn something about themselves (Coles and Timothy 2004). Significant numbers of people from various diasporas travel to their home-lands each year in fulfillment of predictions that heritage tourism is as much related to the individual and social identities of the tourists themselves as it is about the historic places they visit (Breathnach 2006; McIntosh 2008; Poria *et al.* 2003, 2006). Indians and Pakistanis are known for traveling to South Asia from all around the globe, primarily to visit relatives but also to participate in community and socio-religious events. Even within the domestic context, urban orthodox Hindus regularly travel to their ancestral villages to take part in family-related and religious rituals (Hancock 2002). Large numbers of Americans, Canadians, and Australians of Eastern European or Middle Eastern descent travel regularly to their homelands for the same reasons noted above, with sometimes additional motives of demonstrating solidarity with the homeland (e.g., the former communist countries of Eastern Europe and Israel, Palestine, and Lebanon).

African Americans and British, particularly those who have descended from the slave trade, are especially ardent travelers to Africa. For these tourists, the journey is particularly profound but complicated, often wreaking havoc on their emotions and identities as black Americans or British. Many of them seek forgiveness, healing, and closure; others seek revenge and are stirred to anger against the white European and American perpetrators of slavery (Teye and Timothy 2004; Timothy and Teye 2004). Regardless of their experiences in Africa, the visit is nearly always emotionally heavy-laden, confusing, and upsetting in terms of the history of atrocity perpetrated against their ancestors. Many African American tourists are also astonished, disappointed, and hurt when they arrive in Africa expecting to be welcomed to the homeland but instead are ripped off and treated as foreign tourists rather than brothers and sisters (Bruner 1996; Masland 1999).

Living culture is an important part of heritage tourism in the less-developed world. Agricultural landscapes, agrarian lifestyles, arts and handicrafts, villages, languages, musical traditions, spiritual and religious practices, and other elements of the cultural landscape provide much of the appeal for tourism in LDCs. Rice paddies and farming techniques, traditional architecture and building materials, intricate clothing and cloth, exotic-sounding music, vibrant ceremonies, and unusual fragrances and flavors are part of the appeal (Cohen 2001; Gibson and Connell 2005; Hall *et al.* 2003; Howard 2004; Volkman 1990). Often, cultural festivals develop that are based on spiritual traditions, agricultural harvests, or other constituents of culture (Coulon 1999; Hitchcock and Nuryanti 2000; Swearer 1995). Carnival festivals in Trinidad and Brazil, Day of the Dead festivities in Mexico, and Christmas celebrations in Palestine are just a few well-known examples (Brandes 1998; Nurse 2004).

In response to the growing tourist interest in the everyday life of the proletariat, living folk museums have emerged throughout the world and are especially popular among foreign tourists for their claimed "authentic" depictions of daily life (Bruner 2005; Hitchcock *et al.* 1997). Such new developments, however, have received considerable criticism (Ateljevic and Doorne 2005; Hoffstaedter 2008), suggesting that the theatrical setups perpetuate MacCannell's (1973, 1976) notion of staged authenticity, where tourism spaces are rearranged so that cultures, or forms of culture, are "performed" for tourists in order to keep authentic cultural elements away from the view of outsiders and are thus inauthentic, mass-produced misrepresentations of destination cultures.

An interesting and vital part of living culture is culinary heritage, cuisine, and foodways. Among the most favored ethnic or foreign foods associated with dining out in Europe, North America, Australia, and New Zealand are cuisines that originated in the countries of the South. Culinary traditions from Mexico, Thailand, India, the Middle East, West Africa, East Africa, China, Indonesia, and the Caribbean islands are among the most popular in the Western world. The foods, preparatory methods, food-associated rites and rituals, and esthetics are an important part of the heritage product in developing countries (Bessière 2002; Cusack 2000; Hall *et al.* 2003; Tran and Nguyen 1997) because they reflect cultural norms and values, struggles and adaptations to the natural world, the realities of geography and place, refinement through history, intergenerational sharing, and imprints on other aspects of heritage, such as religion and culture.

Historic cities and built heritage are another important resource in the less-developed parts of the world. Built heritage in non-industrialized states can be classified in general terms into two forms: indigenous/native or colonial. Many great and ancient cities have become world-class destinations in Asia, Latin America, Africa, the Middle East, and Eastern Europe. They are significant international gateways and centers of tourism commerce. In most cases, they are composed of indigenous architecture and organic morphology with a substantial mix of colonial influence. Thus, it is not uncommon to see urban spaces incorporating native designs and structural styles with more pragmatic colonial designs. Unfortunately, in many cities, European architecture and urban design have supplanted those of the pre-colonial period.

Archeological sites and ancient monuments are important elements of cultural heritage in LDCs. Often, they are the primary draw, as noted earlier, for international tourists, and their resources, if grandiose enough, can become international icons. Ruins and ancient sites are important components of indigenous culture in locations where material culture was a part of the tangible past.

Other types of heritage resources are important on a worldwide scale but are less prominent in LDCs. For example, industrial heritage has become commonplace in Western Europe, North America, and Australia, owing in part to those regions' transition from fundamentally manufacturing and primary, extractive economies to post-industrial service economies. Thus, remnants of industrializing societies are sometimes now considered things of a distant or recent past, but a past nonetheless, whereas the economies of underdeveloped countries still tend to be highly dependent on extractive (e.g., fishing, mining, logging) activities and heavy industry. Similarly, literary is often geared toward citizens of the developed world, with sites commemorating the lives and writings of famous Western novelists and artists.

Other trends

Other trends have emerged in the developing regions of the world in the realm of heritage tourism. A prominent one today is the notion of pro-poor tourism, or poverty alleviation through tourism. Traditionally, poor citizens have been excluded from planning, policy-making, and development. However, recent calls for more participatory and inclusive forms of tourism development have recognized the need for the poor to benefit from tourism rather than simply bearing the burden of its costs. The Pro-Poor Tourism Partnership (http://www.propoortourism.org.uk) is actively involved in devising and promoting ways in which tourism can be utilized to alleviate poverty and spread tourism income to more sectors of society. Built heritage, living culture, and well-made arts and handicrafts are an important part of these efforts and are recognized as crucial elements of the heritage product upon which communities can base their development efforts.

Another trend deals with United Nations Educational, Scientific, and Cultural Organization (UNESCO). There is a scramble in LDCs to inscribe as many heritage sites as possible on UNESCO's World Heritage List (WHL). As developing countries often have lower levels of global visibility, they frequently use the WHL as a way of making their countries visible. In many cases, this coveted designation is also seen as a way of possibly acquiring international assistance to conserve and manage the sites and a tool for marketing and promotion. In fact, outside of the United States, this UNESCO label is a highly valued promotional tool for developing tourism (Timothy and Boyd 2006b). Unfortunately, there is an erroneous assumption in much of the world that inscription on the WHL will inevitably result in increased visitation and therefore increased tourism earnings (Li *et al.* 2008). While World Heritage Site (WHS) status may in some cases generate additional tourism revenue (Li *et al.* 2008; Yan and Morrison 2007), a tacit assumption that this is the case has so far not been substantiated (Hall and Piggin 2001). Instead, it appears more likely that popular sites will continue to be popular, while less accessible and less popular heritage places will not see considerable growth in arrivals regardless of their WHS designation.

Despite the privilege of being listed by UNESCO, a handful of WHS have come under such human-induced and natural pressure that they have been placed on UNESCO's List of World Heritage in Danger. Several sites have also been de-listed completely because of a lack of proper management and planning. Signatory states are encouraged to eliminate as soon as possible human-induced pressures or risk losing their WHS designation. Table 1.2 illustrates the thirty World Heritage Sites currently on the danger list, all but two of which are in developing countries.

As Table 1.2 denotes, heritage in the developing world faces a variety of threats and challenges. These include, but are not limited to, war and other political conflict, vandalism and human wear, urbanization and agricultural pressures, overcrowding by tourists, and lack of planning and management. While many of these same elements face heritage places in the developed world, many of them are unique to developing countries. The remainder of this book examines these and other issues as they pertain to individual regions and realms of the world.

Table 1.2 UNESCO's list of World Heritage in Danger, July 2008

Country	Name of site	Placed on Danger List	Nature of threat(s)
Afghanistan	Archeological remains of Bamiyan Valley	2003	Intentional destruction, looting, vandalism
Afghanistan	Minaret and archeological remains of Jam	2002	Site deterioration and lack of management
Azerbaijan	Walled city of Baku	2003	Urbanization, natural disasters, lack of conservation policies
Central African Republic	Manovo-Gounda St. Floris National Park	1997	Illegal grazing and poaching
Chile	Humberstone and Santa Laura saltpeter works	2005	Lack of maintenance, looting, vandalism
Côte d'Ivoire	Comoé National Park	2003	Political unrest, poaching, lack of management
Côte d'Ivoire/ Guinea	Mount Nimba Strict Nature Reserve	1992	Mining pressures, refugee settlements
DR Congo	Garamba National Park	1996	Poaching and murder of rangers
DR Congo	Kahuzi-Biega National Park	1997	Illegal timbering, poaching, political unrest, refugee settlement
DR Congo	Okapi Wildlife Reserve	1997	Illegal mining, poaching, armed conflict
DR Congo	Salonga National Park	1999	Poaching, political unrest, destruction of infrastructure
DR Congo	Virunga National Park	1994	Refugee settlements, hunting, poaching, political unrest
Ecuador	Galápagos Islands	2007	Tourism pressures, increased immigration
Egypt	Abu Mena	2001	Agriculture pressures, structural collapse
Ethiopia	Simien National Park	1996	Human settlement, species decline
Germany	Dresden Elbe Valley	2006	Construction, urban pressures
India	Manas Wildlife Sanctuary	1992	Political instability, poaching, species decline
Iran	Bam and its cultural landscape	2004	Natural disasters, lack of conservation
Iraq	Ashur	2003	Lack of protection, potential dam construction
Iraq	Samara archeological city	2007	War and military occupation
Jerusalem	Old City of Jerusalem and its walls	1982	Tourism pressures, urban development, lack of maintenance
Niger	Air and Ténéré Natural Reserves	1992	Political unrest, poaching
Pakistan	Fort and Shalamar Gardens in Lahore	2000	Urban development, lack of maintenance

Table continued next page

Table 1.2 (continued)

Country	Name of site	Placed on Danger List	Nature of threat(s)
Peru	Chan Chan archeological zone	1986	Natural erosion, lack of maintenance
Philippines	Rice terraces of the Philippine Cordilleras	2001	Agriculture pressures, human-induced change
Senegal	Niokola-Koba National Park	2007	Poaching, potential dam construction
Serbia/Kosovo	Medieval monuments in Kosovo	2006	Political instability, lack of maintenance
Tanzania	Ruins of Kilwa Kisiwani and ruins of Songo Mnara	2004	Physical deterioration, lack of maintenance
Venezuela	Coro and its port	2005	Urban development, natural disaster, lack of maintenance
Yemen	Historic town of Zabid	2000	Structural deterioration, lack of maintenance

Source: UNESCO (2008).

This book

The book is divided into two sections. The first section deals with broad conceptual underpinnings, debates, and paradigms in the realm of heritage and heritage tourism that are pertinent to the developing world. The first four chapters, including this introduction, are authored by Dallen Timothy and Gyan Nyaupane. The remaining chapters in Part II are written by experts in the field of heritage tourism, who have considerable research experience and first-hand knowledge about heritage issues in their parts of the world. Unique to this book is its utilization of authors who are either native to the regions they are writing about or who currently live there.

Chapter 2 focuses on the relationship between heritage conservation and tourism, and discusses major challenges and opportunities in heritage conservation in LDCs. The major challenges include financial constraints, ownership issues, agricultural encroachment, looting and illegal digging, colonialism, improper conservation, war and conflict, modernization, heritage overload, lack of cooperation and holistic management, and lack of social and political will. Despite these challenges, heritage conservation provides many opportunities for developing countries. The second part of the chapter shows how heritage-based tourism can be a tool for poverty alleviation, for stimulating the economy, and for generating revenue for heritage conservation and management. The chapter further discusses that, in addition to economic opportunities, heritage tourism can help empower local communities, rejuvenate historic urban spaces, help a country or destination to improve its image, and promote national solidarity. Chapter 3 examines the relationships between politics and heritage by drawing examples from developing countries and using them in political, heritage, and tourism contexts. The chapter discusses some aspects that are prominent in the developing world context including contestation, political uses of the past, power and empowerment, and political instability. Chapter 4 elucidates the negative and positive impacts associated with heritage tourism, which are classified into three domains: physical, social, and economic. The chapter specifically illustrates the consequences of tourism when people's homes, villages, and sacred spaces are open to tourists and concludes that developing countries often give more priority to much-needed revenue generation than to conservation, which typically leads to the deterioration of heritage places.

Part II of the book examines heritage tourism in various developing regions of the world. We use a regional approach, based on the major world realms and regions that are commonly accepted in geography studies today, to discuss heritage and heritage tourism in the developing world for three reasons. First, heritage goes beyond state borders as many developing countries' historical borders were annihilated and rearranged by colonial powers by establishing bigger political units, and deliberately weakening and dismantling indigenous institutions for the sake of control. As an example, the separation of India and Pakistan, and now Kashmir, has been a source of tension between the two countries since their independence. Similarly, many countries have separated because of opposing political ideologies. For example, South Korea and North Korea share the same culture but, in not so distant past, these countries were divided, and Koreans in general have for decades been prohibited from visiting their relatives and heritage across the border.

Second, many countries in the same region share the same history, culture, religion, and politics and, as a result, they face very similar issues related to heritage. This is what unites and binds them as "official" regions (de Blij and Muller 2006). Therefore, the way they tackle these issues can also have common ground among countries in the same region. Finally, heritage tourism is a regional phenomenon, as tourists cross borders to visit temples, stupas, tombs, mosques, churches, and other historical locations. In some instances, one pilgrimage themed route includes more than one country. For example, a Buddhist pilgrim intent on visiting the four most important Buddhist sites must travel to both Nepal and India. Although these sites are located within the same geographic area, they are separated by an international border.

The second section of the book includes nine realms of the developing world as defined by regional geography scholars (de Blij and Muller 2006; Hobbs 2009). We realize, however, that not all countries in the various realms identified here can be considered less-developed countries. On the contrary, most realms and regions include a mix of MDCs and LDCs. For example, Singapore is a well-developed country inside a less-developed region (Southeast Asia), and Japan and South Korea are developed states that share an East Asian regional identity with the less-affluent states of China, Mongolia, and North Korea. Despite this overlap, the regional chapters have been written with entire regions in mind, while attempting to address the issues facing the less-developed constituents of those regions. Each empirical chapter focuses on current issues, opportunities or potential solutions, and unique empirical examples related to heritage tourism in the region. While many scholars include nature as part of the heritage realm, this book focuses primarily on living culture and built environment as the primary components of heritage.

In Chapter 5, Henderson examines heritage tourism issues in Southeast Asia with a focus on socio-cultural, colonial, wartime, and political heritage. The chapter argues that, in addition to its commercial functions, heritage can act as political capital for nation-building. She concludes that, despite some challenges, heritage in Southeast Asia emerges as a core tourism asset that has excellent prospects if it is successfully conserved and sustainably managed with particular risks of neglect, over-exploitation, degradation, and politicization.

Chapter 6 covers the LDCs of East Asia, namely China, Mongolia, and North Korea. In this chapter, Timothy, Wu, and Luvsandavaajav focus on communist–socialist legacies and the region's diversity of cultural landscapes. They demonstrate that, although the three countries' heritage tourism is influenced by their common socialist–communist heritage, the countries demonstrate different levels of heritage connection with state socialism. They also discuss specific challenges each country faces in terms of heritage preservation and management. Owing to the common socialist past of these countries, one of the common issues in the region is a strict form of top-down planning that disallowed essentially all forms of participation in tourism and heritage preservation.

In Chapter 7, Hall covers the countries of the Pacific region. He discusses how two major outside forces, Christianity and colonialism, play a crucial role in changing the cultural identities of the Pacific islands. The chapter further identifies peripherality in the global economy, in terms of geography, commerce, and politics, and threat of global environmental change as two major challenges associated with heritage tourism.

Chapter 8, by Nyaupane and Budruk, discusses South Asian heritage tourism. They discuss how religion and politics have shaped the region's heritage and tourism and outline several issues and opportunities for heritage and tourism that are common to nations of the region in greater depth. These issues include poverty, regional and political conflicts, globalization, and heritage contestation and lack of understanding. Nyaupane and Budruk conclude with recommendations regarding how cultural and ethnic diversity can potentially be turned into an asset that characterizes the region and plays an important role in its economic development.

Chapter 9, by Timothy and Daher, addresses heritage issues in the Middle East (Southwest Asia) and North Africa. They focus on pilgrimage, war and conflict, archeology and empires, and indigenous people. The chapter argues that the terms, "irony" and "dichotomy" help in understanding heritage issues in that part of the globe.

Teye (Chapter 10) examines issues endemic to subSaharan Africa using a triple heritage approach, consisting of traditional or indigenous African heritage, Islamic heritage, and European colonial heritage, in examining tourism in the region. The chapter deals with specific issues, such as colonial influences, heritage identity, and interpretation. Jordan and Duval, in Chapter 11, detail heritage tourism in the Caribbean region with a focus on regulatory and policy environments relevant to the development and maintenance of the past. The chapter also outlines some challenges and critical factors that are key to the success of heritage tourism in the region, which include political will, marketing, capital funding, institutional arrangements, and community participation.

In Chapter 12, Schlüter discusses Latin American heritage and tourism, focusing primarily on a history of political instability and colonialism and how this past influences the region's abundance of indigenous and European heritage. Chapter 13, by Light, Young, and Czepczyński, discusses the heritage and tourism of Central and Eastern Europe in light of the region's communist past and its balancing act existing half in the developed world and half in the less-developed world. Additionally, the authors highlight the emerging identity issues associated with this newly developing region and the choices that are made to demonstrate the past to outsiders.

The book closes with a concluding chapter (Chapter 14) by Nyaupane and Timothy, which draws attention to common challenges faced by developing countries and discusses the implications of heritage tourism and future directions. It is clear that heritage tourism has enormous potential in the less-developed realms of the world. This book aims to shed light on many of the common issues, challenges, and opportunities facing LDCs in their efforts to conserve and exhibit their pasts to tourists.

References

Ateljevic, I. and Doorne, S. (2005) Dialectics of authentication: performing exotic "otherness" in a backpacker enclave of Dali, China. *Journal of Tourism and Cultural Change*, 3(1): 1–17.

Bennet, J. (1993) Travel to Europe ... in the Caribbean. *Europe*, 326: 19–21.

Berger, D.J. (1996) The challenge of integrating Maasai tradition with tourism. In M. F. Price (ed.), *People and Tourism in Fragile Environments*, pp. 175–97. Chichester: Wiley.

Bertocchi, G. and Canova, F. (2002) Did colonization matter for growth? An empirical exploration into the historical causes of Africa's underdevelopment. *European Economic Review*, 46(1): 1851–71.

Bessière, J. (2002) Local development and heritage: traditional food and cuisine as tourist attractions in rural areas. *Sociologia Ruralis*, 38(1): 21–34.

Bhardwaj, S.M. (1983) *Hindu Places of Pilgrimage in India: A Study in Cultural Geography*. Berkeley, CA: University of California Press.

Brandes, S. (1998) The Day of the Dead, Halloween and the quest for Mexican national identity. *Journal of American Folklore*, 11: 359–80.

Breathnach, T. (2006) Looking for the real me: locating the self in heritage tourism. *Journal of Heritage Tourism*, 1(2): 100–20.

Britton, S. (1982) The political economy of tourism in the Third World. *Annals of Tourism Research*, 9: 331–58.

Bruner, E.M. (1996) Tourism in Ghana: the representation of slavery and the return of the Black Diaspora. *American Anthropologist*, 98(2): 290–304.

—— (2005) *Culture on Tour: Ethnographies of Travel*. Chicago, IL: University of Chicago Press.

Cameron, C.M. and Gatewood, J.B. (2008) Beyond sun, sand and sea: the emergent tourism programme in the Turks and Caicos Islands. *Journal of Heritage Tourism*, 3 (1): 55–73.

Cohen, J.H. (2001) Textile, tourism and community development. *Annals of Tourism Research*, 28: 378–98.

Coles, T. and Timothy, D.J. (eds) (2004) *Tourism, Diasporas and Space*. London: Routledge.

Confer, J.C. and Kerstetter, D.L. (2000) Past perfect: explorations of heritage tourism. *Parks and Recreation*, 35(2): 28–38.

Coulon, C. (1999) The Grand Magal in Touba: a religious festival of the Mouride Brotherhood of Senegal. *African Affairs*, 98: 195–210.

Cousineau, P. (2000) *The Art of Pilgrimage: The Seeker's Guide to Making Travel Sacred*. Newbury Port, MA: Red Wheel.

Crosby, A.W. (2004) *Ecological Imperialism: the Biological Expansion of Europe, 900–1900*. Cambridge: Cambridge University Press.

Cusack, I. (2000) African cuisines: recipes for nation-building? *Journal of African Cultural Studies*, 13(2): 207–25.

de Blij, H.J. and Muller, P.O. (2006) *Geography: Realms, Regions and Concepts*, 12th edn. New York: Wiley.

Evans, K. (1998) Competition for heritage space: Cairo's resident/tourist conflict. In D. Tyler, Y. Guerrier and M. Robertson (eds), *Managing Tourism in Cities: Policy, Process and Practice*, pp. 179–92. Chichester: Wiley.

Faye, M.L., McArthur, J.W., Sachs, J.D. and Snow, T. (2004) The challenges facing landlocked developing countries. *Journal of Human Development*, 5(1): 31–68.

Gibson, C. and Connell, J. (2005) *Music and Tourism: On the Road Again*. Clevedon: Channel View.

González, R. and Medina, J. (2003) Cultural tourism and urban management in northwestern Spain: the pilgrimage to Santiago de Compostela. *Tourism Geographies*, 5(4): 446–60.

Hall, C.M. and Piggin, R. (2001) Tourism and World Heritage in OECD countries. *Tourism Recreation Research*, 26(1): 103–5.

Hall, C.M., Sharples, L., Mitchell, R., Macionis, N. and Cambourne, B. (2003) *Food Tourism Around the World: Development, Management and Markets*. Oxford: Butterworth Heinemann.

Hancock, M. (2002) Subjects of heritage in urban southern India. *Environment and Planning D: Society and Space*, 20: 693–717.

Harrison, D. (ed.) (1992) *Tourism and the Less Developed Countries*. London: Belhaven.

Hilton, I. (1989) Shining Path of insurgency. *Geographical Magazine*, 61(8): 22–26.

Hitchcock, M. and Nuryanti, W. (eds) (2000) *Building on Batik: The Globalization of a Craft Community*. Aldershot: Ashgate.

Hitchcock, M., Stanley, N. and Siu, K.C. (1997) The South-east Asian "living museum" and its antecedents. In S. Abram, J. Waldren and D.V.L. MacCleod (eds), *Tourism and Tourism: Identifying with People and Places*, pp. 197–221. Oxford: Berg.

Hobbs, J.J. (2009) *World Regional Geography*, 6th edn. Belmont, CA: Brooks/Cole.

Hoffstaedter, G. (2008) Representing culture in Malaysian cultural theme parks: tensions and contradictions. *Anthropological Forum*, 18(2): 139–60.

Howard, M.C. (2004) A comparative study of the warp ikat patterned textiles of mainland Southeast Asia. *Journal of Cloth and Culture*, 2(2): 176–206.

Huybers, T. (2007) *Tourism in Developing Countries*. London: Edward Elgar.

Irandu, E.M. (2004) The role of tourism in the conservation of cultural heritage in Kenya. *Asia Pacific Journal of Tourism Research*, 9(2): 133–50.

Krakover, S. and Cohen, R. (2001) Visitors and non-visitors to archaeological heritage attractions: the cases of Massada and Avedat, Israel. *Tourism Recreation Research*, 26(1): 27–33.

Landes, D.S. (1998). *The Wealth and Poverty of Nations*. New York: W.W. Norton & Co.

Leung, Y.F. (2001) Environmental impacts of tourism at China's world heritage sites: Huangshan and Chengde. *Tourism Recreation Research*, 26(1): 117–22.

Li, M., Wu, B. and Cai, L. (2008) Tourism development of World Heritage Sites in China: a geographic perspective. *Tourism Management*, 29: 308–19.

Luxner, L. (1999) Reviving Haiti's paradise. *Américas*, 51(4): 48–54.

McCabe, C. (1992) The Caribbean heritage. *Islands*, 12(2): 62–76.

MacCannell, D. (1973) Staged authenticity: arrangements of social space in tourist settings. *American Journal of Sociology*, 79(3): 589–603.

—— (1976) *The Tourist*. New York: Schocken Books.

McIntosh, A.J. (2008) "Back to the future" in heritage and cultural tourism scholarship: a critical lens. Keynote address presented at the conference, Heritage and Cultural Tourism: the Present and Future of the Past, Jerusalem, 17–19 June.

MacKellar, L., Wörgötter, A., and Wörz, J. (2000) *Economic Development Problems of Landlocked Countries*. Vienna: Institute for Advanced Studies.

Masland, T. (1999) "And still I rise!" *Newsweek*, 6 September: 71–72.

Mowforth, M. and Munt, I. (1998) *Tourism and Sustainability: New Tourism in the Third World*. London: Routledge.

New7Wonders (2008) The Official Declaration of the New Seven Wonders of the World. Available from http://www.new7wonders.com/ (accessed August 30, 2008).

Nurse, K. (2004) Trinidad Carnival: festival tourism and cultural industry. *Event Management*, 8(4): 223–30.

Nyaupane, G.P., White, D. and Budruk, M. (2006) Motive-based tourist market segmentation: an application to Native American cultural heritage sites in Arizona, USA. *Journal of Heritage Tourism*, 1(2): 81–99.

Oppermann, M. and Chon, K.S. (1997) *Tourism in Developing Countries*. London: International Thomson Business Press.

Pearce, D.W. and Turner, R.K. (1990) *Economics of Natural Resources and the Environment*. Baltimore, MD: Johns Hopkins University Press.

Population Reference Bureau (2008) *World Population Data Sheet*. Washington, DC: Population Reference Bureau and USAID.

Poria, Y., Biran, A. and Reichel, A. (2006) Tourist perceptions: personal vs. non-personal. *Journal of Heritage Tourism*, 1(2): 121–32.

Poria, Y., Butler, R. and Airey, D. (2003) The core of heritage tourism. *Annals of Tourism Research*, 30: 238–54.

—— (2004) Links between tourists, heritage, and reasons for visiting heritage sites. *Journal of Travel Research*, 43(1): 19–28.

Prentice, R., Davies, A. and Beeho, A. (1997) Seeking generic motivations for visiting and not visiting museums and like cultural attractions. *Museum Management and Curatorship*, 16(1): 45–70.

Sachs, J.D. and Warner, A.M. (1995) *Natural Resource Abundance and Economic Growth*. Cambridge, MA: National Bureau of Economic Research.

Semenov, M.A. and Porter, J.R. (1995) Climatic variability and the modeling of crop yields. *Agricultural and Forest Meteorology*, 73(3/4): 265–83.

Shair, I.M. and Karan, P.P. (1979) Geography of the Islamic pilgrimage. *GeoJournal*, 3 (6): 599–608.

Singh, R.P.B. (2006) Pilgrimage in Hinduism: historical context and modern perspectives. In D.J. Timothy and D.H. Olsen (eds), *Tourism, Religion and Spiritual Journeys*, pp. 220–36. London: Routledge.

Slinger, V. (2000) Ecotourism in the last indigenous Caribbean community. *Annals of Tourism Research*, 27: 520–23.

Swearer, D.K. (1995) *The Buddhist World of Southeast Asia*. Albany: SUNY Press.

Teye, V.B. and Timothy, D.J. (2004) The varied colours of slave heritage in West Africa: white American stakeholders. *Space and Culture*, 7(2): 145–55.

Timothy, D.J. (1997) Tourism and the personal heritage experience. *Annals of Tourism Research*, 34(3): 751–54.

—— (1999) Participatory planning: a view of tourism in Indonesia. *Annals of Tourism Research*, 26(2): 371–91.

Timothy, D.J. and Boyd, S.W. (2003) *Heritage Tourism*. Harlow: Prentice Hall.

—— (2006a) Heritage tourism in the 21st century: valued traditions and new perspectives. *Journal of Heritage Tourism*, 1(1): 1–16.

—— (2006b) World Heritage Sites in the Americas. In A. Leask and A. Fyall (eds), *Managing World Heritage Sites*, pp. 239–49. Oxford: Butterworth Heinemann.

Timothy, D.J. and Teye, V.B. (2004) American children of the African diaspora: journeys to the motherland. In T. Coles and D.J. Timothy (eds), *Tourism, Diasporas and Space*, pp. 111–23. London: Routledge.

Tran, Q.V. and Nguyen, N. (1997) Gastronomic heritage of Vietnam. *Vietnamese Studies*, 125: 25–30.

UNDP (2008) Human Development Report. Available from http://hdr.undp.org/en/media/HDR_20072008_EN_Chapter2.pdf (accessed October 7, 2008).

UNESCO (2008) List of World Heritage Sites in danger. Available from http://whc.unesco.org/en/danger/ (accessed September 15, 2008).

UNWTO (2005) *Cultural Tourism and Poverty Alleviation: The Asia-Pacific Perspective*. Madrid: World Tourism Organization.

—— (2006) *Poverty Alleviation through Tourism: A Compilation of Good Practices*. Madrid: World Tourism Organization.

Volkman, T.A. (1990) Visions and revisions: Toraja culture and the tourist gaze. *American Ethnologist*, 17(1): 91–110.

Wager, J. (1995) Developing a strategy for the Angkor World Heritage Site. *Tourism Management*, 16: 515–23.

Yan, C. and Morrison, A.M. (2007) The influence of visitors' awareness of World Heritage listing: a case study of Huangshan, Xidi and Hongcun in southern Anhui, China. *Journal of Heritage Tourism*, 2(3): 184–95.

2 Protecting the past
Challenges and opportunities

Tourism and protecting cultural heritage

Conserving cultural heritage is as important as conserving the natural environment. Yet, most tourism scholars have focused their discussions of sustainable tourism on the natural world. While some natural realms will in fact recover from the impacts of development and regenerate organically, damaged cultural heritage will not. Built heritage is a non-renewable resource that once destroyed is gone forever. This creates a unique challenge to heritage conservators and managers, who have long had to deal with throngs of tourists clambering on or vandalizing places of historic importance. These impacts will be discussed in greater depth in Chapter 4.

Observers have identified several reasons why heritage is preserved. These include countering the effects of modernization (e.g., demolition of historic structures), building nationalism and preserving collective nostalgia, improving science and education, safeguarding artistic and esthetic values, maintaining environmental diversity, and generating economic value (Timothy and Boyd 2003). While each of these is important in all parts of the world, the final point, economics, is the primary motive for conserving the built and living past in developing regions. Cultural heritage is seen in many places as an economic savior upon which tourism should always be based. Regardless of motive, however, conservation of the historic environment and living culture is critical in today's rapidly modernizing world (Alley 1992), and given what is known about the destructive influences of mass tourism, including mass heritage tourism, heritage protection becomes a more urgent agenda item.

Unfortunately, in the developing world, where much of the earth's magnificent heritage is located, this protection goal is easier said than done. Many challenges exist in underdeveloped regions that often thwart conservation objectives. This chapter examines many of these challenges from socioeconomic, political, and historical perspectives. Because not everything related to heritage protection in the developing world is doom and gloom, the chapter also describes many opportunities that exist for heritage managers and communities in the less affluent parts of the globe.

Challenges

In spite of the importance of conserving heritage for many reasons, including tourism and economic development, there are a number of challenges associated with heritage conservation in the less-developed world. As will be noted in Chapter 4 regarding the negative impacts of heritage-based tourism, many of these issues are also present in the Western or developed world, but they tend to be more pronounced elsewhere.

Financial constraints

Public funding for conservation and preservation is in short supply in the developed world but is even scarcer in less-developed regions. The most glaring problem associated with heritage conservation and management in the developing world is an endemic lack of funds (Zhang 1992). This problem is so severe that it beleaguers public agencies charged with overseeing heritage and hinders many conservation and management efforts (Henson 1989). While community museums have the potential to arouse interest and enthusiasm for a community's heritage (Ronquillo 1992), developing them and conserving heritage are very expensive, and many smaller communities are unable to raise the money to preserve their cultures and artifacts (Zhang 1992). In most parts of the world, capital cities and the home towns of elite leaders are favored in the distribution of conservation budgets, often overlooking the needs of rural areas and small communities.

Insufficient budgets do not allow appropriate public agencies to hire enough guards to protect historic properties from vandals (Ribeiro 1990), or well-trained staff (Rasamuel 1989). Special problems occur in countries where historic properties are located in remote areas; these are often ignored in favor of sites in closer proximity to capital cities—a problem compounded by a lack of funds to transport specialists and materials to marginal regions (Myles 1989).

Because of funding shortages, site managers have had to demonstrate considerable ingenuity in finding ways to support their efforts by sponsoring special events and seeking grants, sponsorships, and donations. As well, most historic sites throughout the developing world (less so in the developed world) charge admission fees, upon which their maintenance and care are dependent (Timothy and Boyd 2003). To allow less affluent residents opportunities to visit their heritage sites, most site managers have adopted a dual pricing system, whereby citizens pay a significantly lower entrance fee than foreign visitors. Besides allowing residents opportunities to visit sites, it also puts the onus upon foreign visitors to finance the sites' operations. Many foreigners, however, find this price differential offensive (Duff-Brown 2001).

On a more global scale, financial crises, such as the one in Asia in the late 1990s and the current crisis in North America, Europe, and Asia, have long-lasting effects on the economies of developing countries (Henderson 1999; Prideaux 1999). Oftentimes, tourism does not provide enough money to sustain heritage protection efforts, and when economies are bad, the problem is compounded further as tourists stay away and as government coffers are depleted on other social and development programs.

Private ownership and human habitation

Most countries face the issue of private ownership. Many historic buildings deemed worthy of conservation are privately owned and, in most cases, the people who own them lack the finances themselves for the upkeep or restoration of their properties. This typically creates ensembles of dilapidated buildings that are unpleasant to look at, let alone live in (Naidu 1994). Such is the case in India where, in a social climate of cynicism regarding favoritism and elitism, the government of India is disinclined to bestow public funds on families and other private owners for the upkeep and conservation of historic buildings (Leech 2004). Many homes and other heritage properties remain in a state of disrepair and continue to degrade via pollution and normal human and natural wear.

Another important issue is human habitation (Castriota 1999). This can be viewed from two primary perspectives. First, many heritage places are overflowing with human tenancy and economic activity. In Leech's (2004: 88) words, "the concept of 'living culture' may be problematic, but it does convey the fact that many of India's centuries-old walled cities are not museums, but living working vibrant cities." Many highly urbanized areas of the developing world face this challenge, as do rural regions, although the pressures of urbanization increase these pressures manifold, particularly in cities that have overrun their carrying capacities. One such example is Jaipur, India, which was originally built for a population of 60,000 but is now inhabited by some 800,000 (Leech 2004).

In the hundreds or even thousands of historic cities where people live in the historic portions of town, the centuries-old buildings are people's homes. It is not only hard for them to fathom why their homes would be of interest to conservationists, let alone tourists, but it also creates a great deal of conflict as governments come in to relocate entire communities so that reconstruction, restoration, or other conservation devices can be implemented in the name of tourism development. Many urban areas, some of which have been designated UNESCO World Heritage Sites, face daily challenges in balancing the needs of a functioning, dynamic city with the need for preservation and conservation (Leech 2004). Good examples of this include the remarkable old cities of Sana'a, Sucre, Salvador de Bahia, Lijiang, Katmandu, Havana, Bam, Baalbek, Lamu, Lima, Istanbul, and dozens more.

Second, individuals and entire villages are known to have established their homes and communities in or around historic structures. This is another major problem in at least three ways. First, they have profound corrosive impacts on oftentimes delicate structures, similar to the way tourists create wear and tear. However, residents may in some cases cause more damage because they live inside the structures full time and utilize them in ways tourists would not.

Building fires, drying clothes, grinding grain, and climbing on edifices and artifacts do irreparable damage. Second, in the course of their daily lives, residents use protected resources (Phillips 1993). They disassemble structures to acquire building materials and utilize artifacts for tools or other devices. While their reasons for doing this are understandable—poverty and a need to survive—it is nonetheless extremely damaging to historic sites and an epidemic problem in much of the world (Chakravarti 2008; Timothy 1999a). Finally, livestock are usually allowed to wander through, graze in, sleep on, and rub against ancient buildings. In places such as India, where cows are sacred and permitted to roam undisturbed freely, some of these problems are intensified (Gourret 1997). Similarly, roaming animals do significant damage to ancient pottery and other small accoutrements.

In most parts of the world, national parks and other protected areas are inhabited by human beings, although there are some reported instances of forced relocations by governments with the establishment of parklands. Often, these protected areas are the ancestral lands of indigenous peoples and therefore are seen as rightfully theirs. This fact, together with increasing land shortages, leads many people to poach, mine, and gather timber illegally off public lands. These are inharmonious with the goals of protected lands and commonly result in arrests, large fines, and conflict between residents and government officials (Aagesen 2000; Phillips 1993; Timothy and Boyd 2006; Ward 1992). Other observers have suggested alternatives, such as sharing park entrance fees with inhabitants, including enclaves in protected areas where locals are allowed to gather and hunt, providing employment opportunities, and encouraging tourism development (Aagesen 2000: 562). Without such compensatory actions, conservation efforts will simply result in local people being left to suffer deprivation and social burdens, bearing costs but receiving few benefits (McLean and Stræde 2003: 513).

Agriculture

Farming is an extremely pressing issue in many heritage places throughout the developing world. The planning documents for several World Heritage Sites, for example, note the threat of encroaching agricultural land use, especially in some countries where mine closures have forced more people to earn a living by farming (Thorsell and Sigaty 2001). In many tropical countries, rainforests are routinely cleared for cattle ranching and other forms of agriculture, affecting both natural and cultural heritage sites, including several on the UNESCO List. Much of the clearing comes dangerously near, and against in some cases, park and preserve boundaries (de Silva and Walker 1998; Timothy and Boyd 2006).

Looting and illegal digging

One of the most salient and urgent concerns in the developing world is looting and illegal digging (Brodie 2003, 2005; Ciochon and James 1989; Lafont 2004; Prott 1996). While this also takes place in developed regions, such as in the United States at Native American sites and in the UK at Roman or Celtic sites, it is especially rife in the less-developed portions of the world for several reasons. First, there is an unfortunate and widespread lack of protective legislation in many countries (Ribeiro 1990), but even where national laws and regulations are in place, they are often ignored or unrecognized by local-level authorities. Similarly, many public overseers are inadequately paid by the state and treat their positions as opportunities to supplement their meager incomes. Thus, for what might be a rather small or large remuneration, a blind eye is often turned away from criminal activities, such as looting, grave-robbing, and illegal digging. Second, fueled by a growing desire in the West for antiquities of various sorts, mostly originating in the developing world, there has been no indication that this trend has slowed down in recent years or will do so in the foreseeable future. This is a multi-billion dollar business, which mostly benefits the wealthy go-betweens and antiquities dealers. In Dempsey's (1994: 23–24) words, it is " ... an underground industry rivaled only by drugs and arms trafficking." Even royal families and high-level officials in developing countries are involved in smuggling antiquities. For example, there was a public outcry against the former King of Nepal, Gyanendra Sah, and the prince for supporting smugglers. Third, because of this growing demand and a dearth of other alternatives, many people see this kind of illicit activity as being more financially lucrative than farming or laboring in some other menial occupation. When families are hungry, parents will do whatever it takes to feed their children, even if it means risking arrest or steep fines. Finally, most archeological sites in developed countries have been well excavated by archeologists for many years, but in the developing world, there are still major projects yet to do and many unexplored areas. Also, antiquities from the "other world" may be more interesting to collectors.

Such issues are particularly noticeable where developing countries and developed countries share the same region. For example, Viking artifacts that originate in Iceland, Denmark, Norway, the UK, or Sweden are extremely rare on today's antiquities market owing to those countries' strict laws and harsh penalties against trade in national heritage (Graham-Campbell 2001). Nearby Russia (and several other areas of Central and Eastern Europe), however, is one of the primary sources of Viking artifacts on the market today, owing to the country's more relaxed laws, inability to enforce existing laws, its sheer size, a flourishing underground economy, a well-established system of organized crime (Politi 1999), and the lower socio-economic status of its general population. Similar examples can be seen in the Middle East. In Israel, artifacts are heavily regulated, although a certain number of more common items are allowed to be sold on the world market. In Syria, Iraq, and Iran, however, illicit trade in antiquities runs rampant (McCalister 2005).

However, there are cases of crackdowns taking place and networks of illegal traders being infiltrated and broken up. In December 1990, for instance, in China, one of the primary sources of ancient artifacts on the world market, 700 cases of tomb robbing were discovered in Shaanxi Province. This allowed the Chinese authorities to prosecute several known criminals and recover more than 2,500 ancient relics (Zhang 1992). In the end, this resulted in tougher laws against illicit excavations and smuggling, but it still runs rampant today throughout China, Southeast Asia, South Asia, Africa, and Latin America (Atwood 2004; Brodie *et al.* 2001).

This problem had become so challenging during the 1960s that UNESCO enacted the Convention on the Means of Prohibiting and Preventing the Illicit Import, Export and Transfer of Ownership of Cultural Property in November 1970. The convention went into force in 1972 and, to date, 116 countries have ratified or accepted the convention, which requires signatory states to enact the following:

- Preventive measures: States are required to monitor trade, take inventories, issue export certificates, impose penal or administrative sanctions, and carry out educational campaigns.
- Restitution provisions: When requested by origin states, state parties will take appropriate steps to recover and return cultural property imported after the entry into force of the convention of both states concerned, as long as the requesting state shall pay adequate compensation to an innocent buyer or person who has valid title to the property.
- International cooperation framework: Cooperation among and between state parties should be developed through the convention. When cultural patrimony is in jeopardy from pillage, the convention provides a mechanism for enacting import and export controls (UNESCO 2008).

Colonialism

Colonialism was in many cases known for attempting to assimilate colonized societies to fit the norms of their European governors. Standardization of language, Christianization of indigenous populations, and subduing many elements of native culture, such as music, dance, celebrations, and traditions, were the hallmarks of these coercions (Hilaire 2003). Indigenous religion and culture were thus replaced by Western belief systems and cultural norms. In this process, many observers believe, the colonial powers broke the indigenous spirit of the people, who have in the intervening years suffered from a sense of cultural loss, misplaced identity, lack of self-determination, low levels of social esteem, and a sense of subjugation (Aziz-al-Ahsan 1998; Mané-Wheoki 1992). Thus, local heritages were suppressed, and in some cases eliminated, in favor of replacement ones. In most cases, the replacements reflected colonial superiority, wealth, and elitist landscapes (Askew 1996). Trotzig (1989) noted that the presence of foreign powers delayed, or even prohibited, an historical indigenous consciousness and identity, although Long (2002) argues that the French restoration efforts in Laos helped lay the foundations for indigenous nationalism. Colonial superpowers sometimes used extreme and brutal means to suppress nationalism and freedom of thought regarding ethnic identity (Adelman 1999;

Loomba 2005). As a result, today much knowledge about some pasts has been lost, and many of the relics that remain in the less-developed world tell much more about European domination than they do about pre-European times.

In countries that were not colonized by European powers, there tends to be a more unified national psyche and heritage than in colonized states. In Bhutan, for instance, a lack of colonial heritage and the country's isolation have allowed the long-term preservation of traditional architecture, culture, and other elements of the past (Lhundup 2002; Nock 1995). With the exception of the country's minority Nepali population, there is a notable unified Bhutanese identity, patriotism, and loyalty to the king and state, and the tradition has been one of protecting heritage over economic growth at all costs (Lhundup 2002: 708).

What few and sometimes insincere efforts were made by the colonizers to preserve the past were often done unsympathetically or with complete disregard for extant social mores and cultural sensitivities. While from their perspective they were doing something good for the indigenes, their efforts were frequently made in ignorance regarding the importance of places and personal heritage among the people. In attempting to preserve the native past, the British, French, Dutch, Spanish, Portuguese, and others museumified elements of culture in ways that were often offensive to the natives. In the process, sacred artifacts and places were defiled; museums took treasures from their natural contexts, secularized them, and framed them in profane space and time (Mané-Wheoki 1992). In the words of Mané-Wheoki (1992: 35), the Western idea of preserving deeply meaningful artifacts in a museum, "far away from the people for whom [they] had originally been carved and built, was tantamount to keeping a brain-dead body artificially alive on a life-support system; tantamount to freezing the corpse; tantamount to placing one's dead grandmother's body on permanent exhibition."

The low priority placed on indigenous culture is evident in the sluggishness associated with the enactment of protective legislation. Most conservation edicts in the developing world were not enacted until the colonials had pulled out and the colonies achieved independence, in many cases as late as the 1970s and 1980s. Ghana's conservation legislation only came into force in 1957, the same year it achieved independence from the British. Ecuador's protective laws were only established in 1979, and Togo's were enacted as recently as the 1990s (Myles 1989; Norton 1989; Sutton 1982). While the Ancient Monuments Preservation Act was passed in India in 1904, only after the nation's independence was it expanded and effected in 1951 (Leech 2004; Ribeiro 1990). The belatedness of these acts has been blamed for much loss of artifacts and archeological sites throughout the less-developed world, particularly as new infrastructures were being constructed, dams were built, and urban expansion was taking place (Henson 1989; Trotzig 1989).

To compound the situation even further, colonial soldiers and their superiors were known to loot indigenous heritage and national treasures for their own wealth accumulation, as well as on behalf of the homeland for museums and royal family collections (Crozier 2000; Evans 1998; Pankhurst 2003).

Improper conservation

Part of the problem associated with colonialism and lack of funding is improper conservation methods. Often during colonial times, indigenous heritage was a low priority, and work on related sites was piecemeal and haphazard. Some of this still remains, but the majority of shoddy work being done today results from budget constraints, inexpert handling of artifacts, and improper restoration techniques—problems that are not necessarily part of the colonial legacy but sometimes are. In the words of Malisius (2003: 37), "as one of the poorest countries in the continent, Bolivia does not at all lack attractions, but competence and capital." Endemic to all developing regions, not just Bolivia, efforts to protect are commonly treated as cosmetic cover-ups rather than structural improvements that will withstand future years and decades of anthropogenic or nature-caused deterioration.

Substandard work in terms of labor and materials utilized is a common problem and was during colonial times as well. Colonial rulers were notorious for inferior preservation techniques and incompatible materials (Timothy and Boyd 2003). Sinking and shifting at Angkor Wat, Cambodia, has caused the temples' walls to lean and damage structural joints. Prior to Cambodia's independence, to reinforce the walls, French supervisors poured concrete pillars and buttresses, often over sculptural reliefs. Iron bands were also wrapped around cracking pillars, which have warped and rusted over the decades. The results are, in the words of Ciochon and James (1989: 56), "ugly and ... irreversible." The French also installed a cement roof over part of the temple complex, when such materials never existed there originally (Ciochon and James 1989). To illustrate this situation, one news magazine notes in the context of Pagan, Myanmar, that

> Almost all of the city's 2,000-odd temples have been fancifully reconstructed, with bright red modern bricks and identical cement finials. In some cases the authorities have built soaring new temples on top of crumbling ancient foundations. In others, they have taken the remains of an original spire and built a new structure to hold it up. This ... verges on Disneyfication ... Undaunted, the junta has devised several schemes to spruce up the ruins, all of which damage the site. A big new road ploughs through the densest cluster of temples, in which a 60-meter viewing platform is under construction nearby ... Similar abuses are under way at Myanmar's other big tourist sites. A 19th century teak palace in Mandalay that burned down during the second world war has been rebuilt in concrete and aluminum ... Since authorities have no idea what it originally looked like, they are simply copying the Mandalay Palace, even though it was built 300 years later ... by a different dynasty.
>
> *Economist* (2004: 65)

Inappropriate cleaning materials are often used as well as part of restoration or preservation efforts. Ciochon and James (1989: 55) provided an example from Angkor, Cambodia, where a team of Indian restorers was hired to clean fungus from the stones of the temple complex. To purge the fungi, the surfaces were treated with sodium pentachlorophenate and zinc silicofluoride. Once cleaned, the workers covered the stones with the sealant polyvinyl acetate to protect the structures from natural elements and recurring fungal spread for up to ten years. Many restoration specialists have criticized the use of these materials, claiming that they actually deteriorate the volatile sandstone, and the polyvinyl acetate is hard to remove. Furthermore, the experts claim that moisture can percolate and build up behind the rubbery synthetic polymer, producing flaking and blistering of stone surfaces. These chemicals, according to Ciochon and James (1989: 56), "produce a splendid effect for five years. It's only after 10 or 15 years that one sees the damage. At first there is yellowing, then the stone opens up cracks. The integrity of the work is seriously damaged."

While many observers chalk these patterns up to incompetence, this may be only part of the problem. There truly are shortages of skilled and qualified staff members who are able to deal with fine restoration techniques and other aspects of conservation (Elkin and Dellino 2001; Goodey 2003; Hill 1990; Lu 2003; Ribeiro 1990; Xiao'an *et al.* 2003; Zhang 1992). Too few university graduates are being produced in areas such as restoration engineering, heritage architecture, and architectural conservation (Leech 2004). Some of this shortage stems from budgetary constraints, as noted earlier, while it also partially results from a lack of understanding of the importance of proper restoration and conservation techniques. Universities in the developing world are finally beginning to realize the importance of their countries' heritage and are beginning to offer courses, majors, and certifications in specialized conservation degrees as a way of remedying part of this problem. In Argentina, there is a nascent realization of the need to protect underwater heritage (e.g., shipwrecks, etc.), so several universities in that country have recently started training specialists in conserving subaquatic cultural resources (Elkin and Dellino 2001).

War and conflict

Some countries are burdened with chronic conflict—civil wars or hostilities between neighbors. These wartime conditions are especially damaging to heritage sites and archeology and cause irreparable damage. War affects heritage in many ways (see Timothy and Boyd 2003). For instance, historic remains are often targeted intentionally by warring factions as a way of destroying morale and injuring the other party's sense of national pride (Talley 1995). Many examples of this exist in the recent wars and other armed conflicts involving Iraq, Afghanistan, Lebanon, the former Yugoslavia, Rwanda, Somalia, Sudan, Sierra Leone, Sri Lanka, and others.

The second effect is heritage as innocent casualty. Here, artifacts are damaged even when they are not targeted directly. The residual effects of battles place heritage sites in danger, such as Angkor Wat during the Khmer

Rouge rebel warfare in Cambodia. Many temples were damaged in those skirmishes, and today walls are still riddled with bullet wounds and the damage done otherwise by the occupying guerillas. In addition, during the Khmer Rouge insurgent occupation of the temples, much of the original wood used for ceilings and joints was burned as firewood (Ciochon and James 1989; Peters 1994). Battles in nearby Laos also destroyed countless temples, stupas, art works, and historical sites (Rattanavong 1994).

Finally, war depletes already indebted economies and takes public monies away from conservation, as well as impeding access to sites that are in need of attention. As already noted, funds are typically very scarce to begin with, and wars always exacerbate the problem. Hornik (1992) asserts that, in many cases, neglect brought on by wars and conflict has a graver negative effect than the direct physical battering suffered during conflict.

Modernization—development versus conservation

The pressures of urban growth pose a significant threat to urban heritage as population and economic pressures mount to expand cities and to construct new buildings in historic districts and rural suburban areas (Castriota 1999; Kaneko 1994; Oren *et al.* 2002). Too often, old buildings are destroyed in the names of modernization and development before the economic justification for saving them has a chance to work (Burton 1993; Long 2002). In the Western world, a common trend is for historic buildings to be renovated and used for modern purposes (e.g., offices, apartments, etc.). In developing regions, however, traditional buildings tend to be razed and replaced anew by Western-style hotels and shopping malls. The expenses associated with pre-serving historic structures often do not justify their maintenance, so they are removed in favor of new buildings that offer more economic promise (Setia-wan and Timothy 2000; Wahyono 1995). In the less-developed world, it is not uncommon for protection of ancient monuments and historic buildings to be viewed as interference in modern development (Sadek 1990). All too often, in some urban areas, indigenous building technology is being lost, and much of it goes unrecorded and evaluated before demolition (Myles 1989; Shackley 1996). While most observers see urbanization and modernization as major barriers to successful heritage protection, Long (2002) submits that the two are not mutually exclusive or completely incompatible, because many of today's modernization programs are based on promoting nationalism and ethnic identity in post-colonial times.

The lack of modernization in Bhutan that is taking place in so much of the less-developed world has allowed for the long-time preservation of traditional and indigenous architecture. Even with strict architectural planning devices in place and rules enforcing building traditions, many traditional building techniques and materials in Thimpu have been replaced by concrete (Nock 1995).

The construction of large dams has become a major threat to heritage places during the past thirty years (Tuna 2000). The Yangtze River was blocked by China's Three Gorges Dam in 2003 to create a large lake for energy production, flood control, irrigation, and public water supply. Unfortunately, the 580-kilometer-long lake being created will cover thousands of years of human history in an area known as a hearth of Chinese culture. However, in the minds of many residents and government officials, the benefits associated with the dam far outweigh the costs (Simons 2003: 50). In a very similar case, the Aswan Dam was erected in the 1960s and 1970s to collect Nile water into a large reservoir. The project was completed in 1970, and Lake Nasser was full by 1976. With the introduction of the dam, major concerns were aired by the international community and UNESCO, because the proposed lake would flood the Nile Valley containing the Abu Simbel temples, one of Egypt's most important markers of ancient civilization. In 1960, a large salvage operation was begun whereby many sites were excavated and documented, and the temples were taken apart and reassembled above the river on dry ground (Luccarelli 1998; Säve-Söderbergh 1987).

Too much of a good thing

One of the most glaring issues facing many countries of the South is their rich array of heritage places and traditions. Tight budgets and a lack of human resources do not allow all, or even a large portion, of the heritage resources to be conserved. In most countries, there is simply too much to conserve. Countries in Asia, Latin America, Southwest Asia, and Africa have "sites of antiquity practically on every corner" (Leech 2004: 88). Thus, decisions have to be made and sites prioritized. Inevitably, this results in many worthwhile sites and artifacts being left to further human-caused and natural decay. A 1999 report by the Indian National Trust for Art and Cultural Heritage listed 213 mosques, 44 temples, 129 residential buildings, 23 palaces, 189 tombs, 117 gateways, and other historical structures totaling more than 1,200 as being worthy and in need of conservation—all just in the city of Delhi alone (Leech 2004: 88).

Lack of cooperation and holistic management

Cooperation is an important principle of sustainable development, because it upholds several principles of sustainability, including efficiency, equity, cultural and ecological integrity, community ownership, integration, holism, balance, and harmony. In the tourism sector, it refers primarily to collaboration between government agencies, private and public sectors, polities that share a common resource, different levels of government within a state, and between private sector businesses/organizations (Timothy 1998). Such collaborative efforts are important for assuring equality of opportunity and the recognition of needs among stakeholders. It also assists in creating harmonious relationships between key players and the environment and economy. The collaborative approach is especially important in the realm of heritage management, for the past is so often contested between groups, and many overlapping authorities lay claim to historic artifacts and places. Without adequate cooperation, sites are neglected, overused, or caught in legal battles, and regional plans are virtually impossible to carry out in an adequate manner.

In most of the world, there is a general lack of holistic management. Each public sector, individual tourism business, or level of government is primarily interested only in its own set of responsibilities. To exacerbate the problem further, agencies and organizations rarely communicate or coordinate their efforts; they carry out their responsibilities without taking into consideration the efforts of other departments, ministries, or the private sector. This can result in overlapping, or parallel, development, ill-fitting projects, over-expended budgets, stalled work, and sectoral fragmentation (Hornik 1992; Timothy 1998, 1999b). Bringing all sides together to achieve the desired results requires coordination among administrations—something that is usually difficult to achieve.

This haphazard situation can be aggravated by the existence of too many agencies involved in issuing permits (Xiao'an *et al.* 2003), taking care of various elements of infrastructure, protecting the environment, interpreting heritage, and marketing tourism. In one case from Indonesia, a provincial agency overseeing telecommunications tore up a major roadway in order to install a series of telephone lines. Once the trenches were filled in and the road repaired, the city agency in charge of sewage treatment re-dug the road a few weeks later (Timothy 1999b). Similarly, many trees, green areas, and parts of the old city wall in Vientiane, Laos, were destroyed to enable a road to be widened. The widening was apparently done without coordination between agencies. Had better coordination occurred, less damage might have been possible (Long 2002).

Lack of social will: poverty and unawareness

Many residents of less-developed regions view preservation with suspicion and ignorance. They equate preservation and conservation with backwardness and see it as antithetical to modernity. This leads to serious problems, as noted earlier, where important historic buildings and ancient monuments are replaced by modern structures, which in most mindsets denotes progress and development. Unlike their counterparts in the West, who tend to value heritage for its sentimental and nostalgic worth, older people in developing nations have few sentimental attachments to historic buildings and other heritage places, because these are too reminiscent of their humble pasts; preservation is often seen as standing still, in opposition to progress, or outmoded (Burton 1993; Myles 1989; Ronquillo 1992; Timothy 1999a). Community members, therefore, pride themselves on constructing new and scrapping the old, and the past is a low priority (Gazaneo 2003).

In the developed world, conservation is often done for esthetic, educational, or other perceived socio-psychological benefits more than for only economic reasons. In less-developed regions, however, the notion of heritage conservation is relatively new and few people appreciate the need for it (Cohen 1978; Henson 1989; Myles 1989). Nearly always, public opinions about heritage are based on its perceived economic value, and there will be little support for it unless residents can connect to it economically (Cohen 1978; Timothy 1999a). This, according to Norton (1989) and Feilden (1993), is understandable because, in places where health care, food, and education are in short supply and where people go hungry every day, conservation of the built environment for conservation's sake is unlikely to be high on their list of priorities.

Lack of political will

In common with individuals, at various levels of government, culture and built heritage are often seen as an unaffordable luxury, when other public services are lacking and money is in short supply. Thus, in many less-developed regions, heritage conservation is not a high priority (Hernández Llosas 2001); in some countries, conservation is done almost solely by foreign investors and international agencies. In Albania, for instance, this was the case in the late 1990s as the country was still recovering from years of strict communist rule. Preserving and conserving heritage took a distant back seat to economic development, which allowed significant looting and other damage to occur in that country. In the words of Aliçka (1997: 78), "the damage that is being inflicted on Albania's culture is perhaps less serious than the economic harm it is suffering, but it is irreparable. In time the economy will recover, but artefacts bearing witness to centuries of culture will have disappeared forever."

Various manifestations of public corruption have already been noted, but the problem is much more widespread than it appears on the surface. In the context of heritage, rules and laws may be disregarded for a fee. One illustration comes from Cambodia, where state officials allowed Hollywood to construct a movie set at Angkor Wat to film the motion picture *Tomb Raider*. Not only was physical damage done to the site, but the images portrayed of Cambodia and Angkor in the movie falsified the nature of Angkor's history and countered the image of high culture the country was attempting to portray to the global community for tourism purposes (Winter 2002). A similar event occurred in Thailand a couple of years earlier, where another motion picture, *The Beach*, was filmed on Phi Phi Leh, a fairly pristine and delicate island in the Andaman Sea. National park conservation laws were broken, with the permission of some high-level officials in Bangkok, to allow much of the natural heritage landscape to be altered dramatically for the movie set, supposedly in the name of promoting tourism to the country. The island's main beach was bulldozed, and many additional, non-native trees and plants were brought in to supplement and replace those already there. There was a huge outcry by environmental groups and the island's villagers, and a general boycott of the movie, suggesting that Thailand's "heritage preservation laws are not for sale." Thailand and Cambodia are not unique in their "selling out" to promises of money in lieu of conserving heritage places.

Even where governments have an interest in preserving heritage and fund it adequately, the focus is all too often solely on built heritage at the expense of preserving living and intangible heritage (e.g., dance, music, languages, traditions) (Turnbull 1998). All the money is pumped into built places because there is a public perception that arts and living culture are not as important or economically valuable.

Opportunities

Despite the challenges discussed above, heritage conservation provides many opportunities. Although tourism and heritage preservation may appear to be strange bedfellows, a synergy can be developed when tourism at heritage sites is properly managed (Engelhardt 2005). "The global megatrend of heritage tourism creates an unprecedented opportunity—particularly in developing countries—for launching and sustaining national and regional conservation strategies, with the backing of the tourism industry" (Ayala 2005: 195). Conservation of heritage ensures that it becomes a resource for development in remote and economically peripheral regions of a country (Greffe 2004). Tourism development at heritage sites can bring improved income and living standards for local people. It stimulates the economy in rural and remote regions by creating demand for agricultural produce, and through infrastructure development projects as it did in Lumbini, Nepal, where many locals were employed as construction workers, and then some were hired as permanent employees to take care of the Buddhist temples.

The effects are not limited only to the local economy. These sites can play an important role as a catalyst for regional economic development as in the case of Vaka Moana in the South Pacific (Ayala 2005). However, one concern about tourism at historic places is that many sites are located in isolated areas where tourists visit as day-trippers and spend very little money at the site or in surrounding areas (Nyaupane 2008). The best way to deal with this issue is to increase tourists' length of stay through adding value to heritage artifacts. Ayala (2005) explained a three-step value-adding process, which may be beneficial for the long-term success and sustainability of heritage tourism. The first step is identifying the heritage product and where it fits within the natural and cultural theme of the destination at the scale of a country or region. The second step is to equip the heritage with conservation guarantees. To succeed, the tourism industry depends on remnants of the past, which is a good reason for them to help preserve heritage. The third step is diversification and upgrading of the heritage experience. "A lively local culture of dance, theater, poetry, painting and even food, all encourage a visitor to stay longer in the area and give the visitor greater depth of understanding about traditional local culture" (Engelhardt 2005: 181). The heritage experience can therefore be enhanced by interpretation (Ayala 2005). Further, to make tourism a viable tool for economic development and cultural preservation, heritage planners and managers have to address the following issues:

- Information for potential tourists.
- Quality (authenticity) of tourism products and sites (interpretation).
- Conservation and management of sites with respect for a site's carrying capacity. This will require the tourism industry to cooperate with and work under the guidance of professional conservators.
- Financing, so that the increased needs of the sites in terms of maintenance and presentation, which tourism demands, are able to be met from the profit revenues of the tourism industry, not from dwindling public funds.
- Endogenous planning, indigenous management, and profit sharing by the affected local community (Engelhardt 2005: 178).

Heritage tourism also helps empower local communities, as can be seen in Hue, the former capital of Vietnam (Engelhardt 2005). Three major concerns have to be addressed for local empowerment to be realized, namely finance, zoning, and integrating living and historical cultures (Engelhardt 2005). To sustain the requisite financial resources for heritage conservation, a long-term development master plan is required, one that incorporates both preservation and development concerns. Innovative public–private partnerships should also be established to link conservation efforts to sustainable tourism development at heritage sites. Similarly, zoning and carrying capacity have to be implemented to protect fragile ruins, archeological remains, monuments, and buildings.

Heritage conservation rejuvenates historic urban space through renovation, restoration, and reconstruction of historic buildings. This makes urban spaces more livable and attractive for investment. Historic preservation may help achieve sustainable economic growth (Munasinghe 2005). With the new historic appearance of the city, along with cultural activities and attractions, the city can attract tourists, expand its services sector, provide facilities such as parking, and help maintain roads, security, and public transportation (Munasinghe 2005). In the city of Pelourinho, Brazil's original capital, many positive changes to the historic environment have been attributed to a growth in tourism, including renovations to historic structures, but also in terms of the social environment by reducing drug use and prostitution in the historic center (Galanternick 1999). Likewise, negative images can be positively sold under the heritage theme; for example, redundant coal-mining infrastructure, war-torn towns and buildings, and waterfronts can be offered as part of the heritage product (Prentice 2005).

Maintenance and preservation of cultural heritage can help create awareness of, and pride in, history and civilization (Prentice 2005: 177). It creates pride in communities for their heritage and provides self-confidence in their culture and heritage, which can result in more local efforts to protect the cultural past. "In the last decade of the nineteenth century, a slow but ever-growing interest developed in the rediscovery of the historic values and architectural significance of the Spanish colonial heritage. This was accompanied later by a critical evaluation of the restoration activities that had occurred" (Gazaneo 2003: 414).

Heritage preservation also provides avenues for different stakeholders to open dialog and cooperate. When local heritage becomes a center of attraction to international visitors, remote areas receive more attention from national governments and the international community, thereby reducing their peripherality from the capital's perspective. Ignored and neglected regions can soon become nationally and internationally recognized sites (Leask 2006).

Heritage is a source and symbol of identity. Both tangible and intangible heritage plays an important role in creating individual, community, and national identity. Castells (2000) divided individual identity into two types: "defense identity" and "project identity." Defense identity is reactive as local arts, history, and traditions are reinforced for the protection of heritage from globalization and other threats. Project identity, on the other hand, goes beyond local place. Each individual can have multiple identities, and many travel to various places to incorporate different cultures into their own identity. In this type of identity, intangible heritage is transported from one place to another without losing all semblances of authenticity (Gonzalez 2008). For nations, heritage is a means of affirming their national identity and promoting solidarity (Greffe 2004; Henderson 2001). There are opportunities for national unity and global recognition through the World Heritage List. Individual and group identity is "a multi-faceted concept that encompasses attitudes, values, traditions, heritage and myths (Henderson 2001: 220). These factors help bind individuals and groups together and provide a sense of belonging as members of a given society (Henderson 2001). Such conditions resemble Scheyvens' (1999) social and psychological forms of empowerment, wherein communities take greater pride in their heritage and a mutual sense of community self-esteem and solidarity are created. Identity cannot be easily observed; rather, it can only be manifested in different forms of expression, of which heritage is one.

Conclusion

This chapter discussed the challenges and opportunities of preserving the past in developing countries. Although both developed and developing countries are struggling with the preservation of the past, developing countries appear to be facing more challenges and more unique obstacles. Many of these challenges are economic in nature. The protection and renovation of historic buildings, tombs, and temples is an expensive endeavor. For developing countries, where the majority of the population is struggling to feed itself, heritage preservation is often not a high priority. There is a budget scarcity for policing and maintaining valuable heritage and cultural assets. As a result, governments and communities in less-developed countries must rely on foreign investments and donor agencies. The limited budget allocations and foreign aid are often spent improperly because of widespread corruption.

In many cases, heritage not only goes unpreserved in museums, but ancient arts and artifacts are showcased in people's private homes and farms, in temples and public squares, and in open fields, which are heavily used by people and animals. Additionally, many less-developed countries have recently experienced and are currently experiencing political instability, civil wars, and other conflict, during which historic remains are targeted to destroy the identity of others. In addition to conflict, there is also a lack of protective legislation in many developing countries, and even that what does exist is frequently not enforced. Although many developing countries colonized by Europeans have started establishing legislative frameworks to preserve the pre-colonial past, many historical items and locations were lost before independence was granted.

Nonetheless, if properly managed, heritage preservation can help uplift the economy of remote and neglected regions of developing countries. Heritage tourism has the ability to help diversify agriculture-based economies and stimulates the growth of other economic sectors. As heritage-based tourism has continued to increase in popularity, communities and regions where historic sites are located have receive increased attention from national governments and international agencies. In some cases, this has resulted in additional development opportunities and improved quality of life through such innovations as roads, airports, and schools.

Many heritage sites in developing countries collect entrance fees from tourists, which provide conservation funding. Less-developed countries that are culturally rich have a distinct competitive advantage over developed countries. Heritage conservation can rejuvenate historic urban spaces through restoration, reconstruction, and conservation (Setiawan and Timothy 2000). For many developing countries, heritage can be a symbol of pride and national identity. The presence of world heritage also provides an opportunity to receive international support. For example, Nepal has received considerable financial support from Japan to preserve and manage Lumbini, the birthplace of Buddha.

The role of UNESCO can be crucial in preserving the cultural heritage of less-developed countries, where governments have few technical and financial resources for conservation. UNESCO's approach is "to promote the development of cultural tourism, not as an end itself, but as a tool for the preservation and enhancement of a society's culture, its physical and intangible heritage, and its environment" (Prentice 2005: 177). UNESCO's World Heritage program has created awareness about preserving heritage sites in developing countries, as it is assumed that being on the List elevates the status of the destination.

As this chapter has highlighted, many challenges lie in the way of developing heritage-based tourism in the less-developed world. Nonetheless, there are an equal number of positive opportunities. The growing recognition of the Third World's heritage resources by UNESCO and other agencies is indicative of that realization. Sound planning in accordance with the principles of sustainable development will allow cultural heritage to continue to be the foundation for desirable tourism experiences in developing countries far into the future.

References

Aagesen, D. (2000) Rights to land and resources in Argentina's Alerces National Park. *Bulletin of Latin American Research*, 19: 547–69.

Adelman, J. (1999) *Colonial Legacies: The Problem of Persistence in Latin American History.* London: Routledge.

Aliçka, Y. (1997) Albania's threatened heritage. *The UNESCO Courier*, July/August: 78–79.

Alley, K.D. (1992) Heritage conservation and urban development in India. *Practicing Anthropology*, 14(2): 23–26.

Askew, M. (1996) The rise of Moradok and the decline of the Yarn: heritage and cultural construction in urban Thailand. *Sojourn*, 11(2): 183–210.

Atwood, R. (2004) *Stealing History: Tomb Raiders, Smugglers, and the Looting of the Ancient World.* New York: St. Martin's Press.

Ayala, H. (2005). Vaka Moana—a road map for the South Pacific economy. In A. Hooper (ed.), *Culture and Sustainable Development in the Pacific*, pp. 190–206. Canberra: Asia Pacific Press at the Australian National University.

Aziz-al-Ahsan, S. (1998) Muslim political culture, colonial heritage, and regime transformation: South Asia and the Arab Middle East. *Journal of South Asian and Middle Eastern Studies*, 21(2): 1–20.

Brodie, N. (2003) Stolen history: looting and illicit trade. *Museum International*, 55(3/4): 10–22.

—— (2005) Illicit antiquities: the theft of culture. In G. Corsane (ed.), *Heritage, Museums and Galleries: An Introductory Reader*, pp.122–40. London: Routledge.

Brodie, N., Doole, J. and Renfrew, C. (eds) (2001) *Trade in Illicit Antiquities.* Cambridge: McDonald Institute.

Burton, S. (1993) History with a bottom line. *Time*, 12 July: 36–37.

Castells, M. (2000). *La era de la informacion. La sociedata red.* Madrid: Alianza Editorial.

Castriota, L. (1999) Living in a World Heritage Site: preservation policies and local history in Ouro Preto, Brazil. *Traditional Dwellings and Settlements Review*, 10(2): 7–19.

Chakravarti, I. (2008) Heritage tourism and community participation: a case study of the Sindhudurg Fort, India. In B. Prideaux, D.J. Timothy and K.S. Chon (eds), *Cultural and Heritage Tourism in Asia and the Pacific*, pp. 189–202. London: Routledge.

Ciochon, R. and James, J. (1989) The battle of Angkor Wat. *New Scientist*, 124: 52–57.

Cohen, E. (1978) The impact of tourism on the physical environment. *Annals of Tourism Research*, 5: 215–37.

Crozier, B. (2000) From earliest contacts: an examination of Inuit and Aleut art in Scottish collections. In M. Hitchcock and K. Teague (eds), *Souvenirs: The Material Culture of Tourism*, pp. 52–71. Aldershot: Avebury.

Dempsey, M. (1994) Protectors of Peru's shining past. *New Scientist*, 20 August: 23–25.

de Silva, A. and Walker, B. (1998) Saving Sri Lanka's living heritage. *Biologist*, 45(2): 71–75.

Duff-Brown, B. (2001) Taj Mahal tickets costly: foreign tourists refusing to pay, hurting business. *Arizona Republic*, 3 September: A27.

Economist (2004) Among the ruins. *Economist*, 28 February: 65.

Elkin, D. and Dellino, V. (2001) Underwater heritage: the case of Argentina. *Bulletin of the Australian Institute for Maritime Archaeology*, 25: 89–96.

Engelhardt, R.A. (2005) Culturally and ecologically sustainable tourism development through local community management. In A. Hooper (ed.), *Culture and Sustainable*

Development in the Pacific, pp. 174–89. Canberra: Asia Pacific Press at the Australian National University.

Evans, G. (1998) Mementoes to take home: the ancient trade in souvenirs. In J.M. Fladmark (ed.), *In Search of Heritage: As Pilgrim or Tourist?*, pp. 105–26. Shaftesbury: Donhead.

Feilden, B.M. (1993) Is conservation of cultural heritage relevant to South Asia? *South Asian Studies*, 9: 1–10.

Galanternick, M. (1999) Real bargain. *Latin Trade*, 7(8): 27–28.

Gazaneo, J.O. (2003) A quest for preservation … for what identity? *The Journal of Architecture*, 8: 411–19.

Gonzalez, M.V. (2008) Intangible heritage tourism and identity. *Tourism Management*, 29(4): 807–10.

Goodey, B. (2003) Interpretive planning in a historic urban context: the case of Porto Seguro, Brazil. *Urban Design International*, 8: 85–94.

Gourret, L. (1997) Taxila: the cradle of Gandhara art. *The UNESCO Courier*, October: 42–44.

Graham-Campbell, J. (2001) *The Viking World*. London: Francis Lincoln.

Greffe, X. (2004) Is heritage an asset or a liability? *Journal of Cultural Heritage*, 5(3): 301–9.

Guo, C. (2006) Tourism and the spiritual philosophies of the "Orient". In D.J. Timothy and D.H. Olsen (eds), *Tourism, Religion and Spiritual Journeys*, pp. 121–38. London: Routledge.

Henderson, J. (2001) Heritage, identity and tourism in Hong Kong. *International Journal of Heritage Studies*, 7(3): 219–35.

Henderson, J.C. (1999) Southeast Asian tourism and the financial crisis: Indonesia and Thailand compared. *Current Issues in Tourism*, 2(4): 294–303.

Henson, F.G. (1989) Historical development and attendant problems of cultural resource management in the Philippines. In H. Cleere (ed.), *Archaeological Heritage Management in the Modern World*, pp. 109–17. London: Unwin Hyman.

Hernández Llosas, M.I. (2001) Archaeology, management and potential cultural tourism in Andean Argentina (South America): the case of Quebrada de Humahuaca. *Bulletin of the Australian Institute for Maritime Archaeology*, 25: 97–108.

Hilaire, J. (2003) Haiti: the Creole heritage today. In G. Collier and U. Fleischmann (eds), *A Pepper-Pot of Cultures: Aspects of Creolization in the Caribbean*, pp. 199–212. Amsterdam: Rodopi.

Hill, C. (1990) The paradox of tourism in Costa Rica. *Cultural Survival Quarterly*, 14 (1): 14–19

Hornik, R. (1992) The battle of Angkor. *Time*, 6 April: 54–56.

Kaneko, K. (1994) The safeguarding and promotion of Vietnamese ethno-forms: protecting the intangible cultural heritage of minority groups in Vietnam. *Vietnamese Studies*, 112: 44–50.

Lafont, M. (2004) *Pillaging Cambodia: The Illicit Traffic in Khmer Art*. Jefferson, NC: McFarland and Co.

Leask, A. (2006) World Heritage Site designation. In A. Leask and A. Flall (eds), *Managing World Heritage Sites*. New York: Butterworth Heinemann.

Leech, K. (2004) The ravage of India. *Blueprint*, 218: 86–91.

Lhundup, S. (2002) The genesis of environmental ethics and sustaining its heritage in the Kingdom of Bhutan. *Georgetown International Environmental Law Review*, 14 (4): 693–740.

Long, C. (2002) A history of urban planning policy and heritage protection in Vientiane, Laos. *International Development Planning Review*, 24(2): 127–44.

Loomba, A. (2005) *Colonialism/Postcolonialism*. London: Routledge.

Lu, J. (2003) Informationizing China's cultural heritage: status quo, problems and countermeasures. *Social Sciences in China*, 24(1): 144–50.

Luccarelli, L. (1998) World heritage in China. *China Today*, 47(8): 59–61.

McCalister, A. (2005) Organized crime and the theft of Iraqi antiquities. *Trends in Organized Crime*, 9(1): 23–37.

McLean, J. and Stræde, S. (2003) Conservation, relocation, and the paradigms of park and people management—a case study of Padampur Villages and the Royal Chitwan National Park, Nepal. *Society and Natural Resources*, 16: 509–26.

Malisius, U. (2003) Tourism development with obstacles: the case of Bolivia. *Trialog*, 79: 32–37.

Mané-Wheoki, J. (1992) Sacred sites, heritage and conservation: differing perspectives on cultural significance in the South Pacific. *Historic Environment*, 9(1/2): 32–36.

Mowforth, M. and Munt, I. (1998) *Tourism and Sustainability: New Tourism in the Third World*. London: Routledge.

Munasinghe, H. (2005). The politics of the past: constructing a national identity through heritage conservation. *International Journal of Heritage Studies*, 11(3), 251–60.

Myles, K. (1989) Cultural resource management in Sub-Saharan Africa: Nigeria, Togo and Ghana. In H. Cleere (ed.), *Archaeological Heritage Management in the Modern World*, pp. 118–27. London: Unwin Hyman.

Naidu, R. (1994) A conceptual framework for the renewal of walled cities in India. *Ekistics*, 61: 298–305.

Nock, D. (1995) The architecture of Bhutan. *The Architectural Review*, 198: 78–81.

Norton, P. (1989) Archaeological rescue and conservation in the north Andean area. In H. Cleere (ed.) *Archaeological Heritage Management in the Modern World*, pp. 142–45. London: Unwin Hyman.

Nyaupane, G.P. (in press) Heritage complexity and tourism: the case of Lumbini, Nepal. *Journal of Heritage Tourism*.

Oren, U., Woodcock, D.G. and Var, T. (2002) Sustaining tourism development: a case of Cumalikizik, Turkey. *Tourism Analysis*, 6: 253–57.

Pankhurst, R. (2003) Ethiopia and the return of Africa's cultural heritage. *Africa Quarterly*, 43(3): 27–39.

Peters, H.A. (1994) Cultural heritage at the National Museum of Cambodia. *Oriental Art*, 40(1): 20–27.

Phillips, D. (1993) Planning in the third world: a conservation project in Vietnam. *Australian Parks and Recreation*, 29(3): 14–17.

Politi, A. (1999) The new dimensions of organized crime in Southeastern Europe. *The International Spectator,* 34(3): 49–58.

Prentice, R. (2005) Heritage: a key sector in the "new" tourism. In G. Corsane (ed.), *Heritage, Museums and Galleries: An Introductory Reader*, pp. 243–56. New York: Routledge.

Prideaux, B. (1999) Tourism perspectives of the Asian financial crisis: lessons for the future. *Current Issues in Tourism*, 2(4): 279–93.

Prott, L.V. (1996) Saving the heritage: UNESCO's action against illicit traffic in Africa. In P.R. Schmidt and R.J. McIntosh (eds), *Plundering Africa's Past*, pp. 29–44. Bloomington, IN: Indiana University Press.

Rasamuel, D. (1989) Problems in the conservation and restoration of ruined buildings in Madagascar. In H. Cleere (ed.), *Archaeological Heritage Management in the Modern World*, pp. 128–41. London: Unwin Hyman.

Rattanavong, H. (1994) To preserve the cultural heritage of the multi-ethnic people of Laos. *Vietnamese Studies*, 112: 107–11.

Ribeiro, E.F.N. (1990) The existing and emerging framework for heritage conservation in India. *Third World Planning Review*, 12(4): 338–43.

Ronquillo, W.P. (1992) Community museums in the Philippines: their potential in enhancing the natural and cultural heritage of the country. *Philippine Quarterly of Culture and Society*, 20: 317–23.

Sadek, H. (1990) Treasures of ancient Egypt. *National Parks*, 64(5): 16–17.

Säve-Söderbergh, T. (1987) *Temples and Tombs of Ancient Nubia: The International Rescue Campaign at Abu Simbel, Philae and Other Sites*. Paris: UNESCO.

Scheyvens, R. (1999) Ecotourism and the empowerment of local communities. *Tourism Management*, 20: 245–49.

Setiawan, B. and Timothy, D.J. (2000) Existing urban management frameworks and heritage conservation in Indonesia. *Asia Pacific Journal of Tourism Research*, 5(2): 76–79.

Shackley, M. (1996) Too much room at the inn? *Annals of Tourism Research*, 23: 449–62.

Simons, C. (2003) Before the flood. *Far Eastern Economic Review*, 166(16): 48–51.

Sutton, M.D. (1982) Interpretation around the world. In G.W. Sharpe (ed.), *Interpreting the Environment*, pp. 644–63. New York: Wiley.

Talley, M.K. (1995) The old road and the mind's internal heaven: preservation of the cultural heritage in times of armed conflict. *Museum Management and Curatorship*, 14(1): 57–64.

Thorsell, J. and Sigaty, T. (2001) Human use in World Heritage natural sites: a global inventory. *Tourism Recreation Research*, 26(1): 85–101.

Timothy, D.J. (1998) Cooperative tourism planning in a developing destination. *Journal of Sustainable Tourism*, 6(1): 52–68.

—— (1999a) Built heritage, tourism and conservation in developing countries: challenges and opportunities. *Journal of Tourism*, 4: 5–17.

—— (1999b) Participatory planning: a view of tourism in Indonesia. *Annals of Tourism Research*, 26(2): 371–91.

Timothy, D.J. and Boyd, S.W. (2003) *Heritage Tourism*. Harlow: Prentice Hall.

—— (2006) World Heritage Sites in the Americas. In A. Leask and A. Fyall (eds), *Managing World Heritage Sites*, pp. 235–45. Oxford: Butterworth Heinemann.

Trotzig, G. (1989) The "cultural dimension of development"—an archaeological approach. In H. Cleere (ed.), *Archaeological Heritage Management in the Modern World*, pp. 59–63. London: Unwin Hyman.

Tuna, N. (2000) Preserving Turkey's cultural heritage at reservoir sites. *Hydropower and Dams*, 7(6): 42–46.

Turnbull, R. (1998) A fragile dance. *Far Eastern Economic Review*, 161(52): 36–38.

UNESCO (2008) Records of the General Conference. Available from http://unesdoc. unesco.org/images/0011/001140/114046e.pdf#page = 130 (accessed 30 July 2008).

Wahyono, H. (1995) Developing an Historic Preservation Area as a tourist attraction: the Old City of Semarang, Indonesia. Unpublished Master's thesis, School of Urban and Regional Planning, University of Waterloo, Canada.

Ward, G.C. (1992) India's wildlife dilemma. *National Geographic*, 181(5): 2–28.

Winter, T. (2002) Angkor meets *Tomb Raider*: setting the scene. *International Journal of Heritage Studies*, 8(4): 323–36.

Xiao'an, W., Qun, D. and Decheng, P. (2003) On sustainable tourism of China's world-class cultural and natural heritage sites. *Social Sciences in China*, 24(1): 160–68.

Zhang, D. (1992) Protecting China's rich heritage of cultural relics. *China Today*, 41 (6): 14–17.

3 The politics of heritage

Introduction

Heritage is not just the past but also a representation or a reinterpretation of the past (Graham *et al.* 2000; Lowenthal 1997). The legacies and relics inherited from the past are not randomly preserved, but they are selectively chosen and have survived many political upheavals, and been shaped and reshaped by the politics of the past and present. Thus, heritage is inherently a political entity. It is not only influenced by politics, but it also influences politics. Heritage can be a foundation of nation-building but also a source of civil war. Although many of the relationships between heritage and politics are historical in nature, there are some countries, such as some in Eastern Europe, where heritage and its meaning are in transition as one political ideology has been replaced by another.

This chapter seeks to explore the relationship between politics and heritage by drawing on examples from developing countries and using them in political, heritage, and tourism contexts. Many aspects of heritage politics discussed in this chapter also apply to the Western, industrial world; however, more focus has been placed on situations and examples from developing countries. In particular, the chapter focuses on concepts that are more prominent in the developing world context, including heritage contestation, political uses of the past, power and empowerment, and political instability.

Multiple heritages and contestation

History has a way of creating contentious heritage situations. Each view of history and how it is presented is subject to discordance and disagreement (Tunbridge and Ashworth 1996), and of course all views claim to be accurate representations of the past. Several types of contested heritage exist. Olsen and Timothy (2002) outlined three types of contestation. The first is where different social groups claim the same heritage places, events, and artifacts, but their versions of what took place differ. In this instance, one group's heritage may be replaced entirely by that of another group. In most cases, each party claims its own "objective truth" about the past, often demeaning, or disinheriting, opposing views of the same location or event. In the Middle East, Jerusalem is a very good example of this, where multiple ethnicities, religions, and societies lay claim to the same heritage places (Timothy and Emmett forthcoming).

Another oft-cited example is the Chinese (24 percent) and Malay (50 percent) populations of Malaysia. In addition to the indigenous Malay peoples, waves of Chinese immigration over the centuries have played a critical role in the development of the country's cities, including language use, cultural traditions, religion, urban morphology, architecture, cuisine, and commerce. Today, the Chinese minority is the second largest ethnic group and comprises nearly one quarter of the population. However, cultural policies and intentional construction of an idealized Malay past in Malaysia de-emphasizes the Chinese population's role in the creation of the modern state (Cartier 1996). This intentional writing out of history, or at least diminishing it, is manifested as societal amnesia (discussed below) and in nationalist policies that emphasize one parallel heritage over another. According to Worden (2001: 210), the national culture of Malaysia must be, by official declaration, based on an "indigenous culture of this region." Thus, the Chinese past is relegated to the national periphery and de-emphasized in favor of the majority Malays.

The second variation of contestation is when heritage is interpreted and used differently by various divisions within a single group, such as a national population or a religion. Occasionally, subgroups within broader groups will interpret their mutual heritage differently, also resulting in contestation and disinheritance. On a small scale, the Church of the Holy Sepulcher in Jerusalem, built over the location where many Orthodox and Roman Catholics believe Jesus Christ was crucified and laid to rest, is composed of a successively built microcosmic labyrinth of overlapping claims of ownership and disputed territorial entitlements (Emmett 2000). This has resulted in power struggles, protectionist conflicts, and stalemates between sects of the broader Christian faith.

Colonialism lies at the root of some of this type of contention. When political boundaries were delimited by European metropoles for their administrative convenience, often with little socio-cultural or historical rationale, many tribes and ethnicities were divided, while others were placed together in loosely defined states that have since struggled to create and maintain a unified national identity. This condition of imposed statehood has resulted in many cultural clashes, identity crises among various ethnic groups, and disinherited pasts in favor of the dominant ethnicities in power. In Vietnam, for example, there are more than fifty ethnic groups vying for recognition as important elements of a national heritage and tourism industry (Kaneko 1994; Thinh 1999).

The final category of heritage contestation occurs within the context of parallel pasts, or when more than one history occurs at the same place and time. In 1992, the 400-year-old Babari Masjid in Ayodhya, India, the birthplace of Lord Rama, was destroyed by Hindu nationalists, because the Hindus claimed it was built unlawfully on the site of an ancient temple dedicated to Rama. Their plan was to rebuild the temple on the site of the Mosque. This led to severe riots and violent clashes between Muslim and Hindu Indians far into the new millennium and has further soured relations between Pakistan and India (Leech 2004).

Likewise, colonial and indigenous histories often unfolded side by side and are often viewed and presented quite differently (Adams 2003; Hancock 2002). In this case, the heritage of the people in power is emphasized in educational curricula, national identity exercises, and tourism (Timothy 2007b). These dilemmas are a constant challenge in Africa and Asia, and bodies responsible for heritage conservation and tourism development are faced with crises related to which heritage to present and which truths should be interpreted as official state doctrine (Timothy and Boyd 2003).

Similar conditions have plagued various parts of Africa, owing to that continent's "triple cultural heritage." According to Ghana's first president, Nkrumah, post-colonial Africa inherited three heritages: indigenous traditions, Islamic traditions, and cultural elements from Western religious and secular civilizations (Keita 1988). Nkrumah also argued that Africa must come to terms with this tripartite heritage and seek to accommodate all of them or face internal upheaval.

Political uses of the past

Being inherently about power and control, heritage is often utilized intentionally by governments to achieve some measured ends and to demonstrate their authority over people and places. There are multitudes of uses of heritage and manipulations of tourism, but the most common and influential of these have been outlined by Kim *et al.* (2007), Timothy (2007b), and Timothy and Boyd (2003), and all are related to heritage tourism in one way or another. One of the most common political uses of tourism is travel warnings. These periodic statements by governments warning their citizens against travel to certain other countries is a legitimate way to caution against potentially risky situations abroad, but it can also carry political undercurrents as a mechanism to control where citizens visit and where their money is spent. Most Western countries issue travel warnings to their citizens for a variety of reasons, including political instability, natural disasters, public protests, health concerns, war, and threats of terrorism and kidnapping. Such official warnings exert great power on Western tourists in their decision-making processes and in spreading apprehension about travel in general and specific destinations (Henderson 2003; Löwenheim 2007; Thapa 2003).

Similar to travel warnings, outright bans on travel can be a retaliatory tool to punish countries that do not comply with another country's demands or with the world community's normative policies. The US government has in the past enacted bans or embargos as a result of negative relations and in an effort to castigate "rogue" states that did not live up to US policies or that demonstrated various degrees of aggression toward the US. The current economic ban against Cuba, which affects travel by US citizens, is a good example (Weinmann 2004). Tourism is thriving in Cuba with approximately two million tourists from Canada, Europe, and Latin America visiting the island state each year (Henthorne and Miller 2003; Jayawardena 2003). Americans comprise a substantial portion of that number as well (more than 100,000 annually), despite the US economic embargo (Jayawardena 2003), which actually limits American spending on the island more than it outright forbids travel there. Many observers thus claim the law to be a failure. Similar US restrictions have existed in the past for North Korea, Cambodia, and Vietnam, although these have now been lifted (Mowry 1999). Aside from the obvious deterring of people from visiting heritage places, in one sense, these restrictions are part of a heritage narrative themselves, at least in the sense that they are based on historical relations between states and events that have been or are currently taking place in the countries on the warning list.

A third political exploitation of tourism occurs when a state or group of states dangle it before another state to entice, to gain favor, or garner support for one view of an international issue. The promise of increased arrivals and foreign support is thus a bribe of sorts to encourage compliance or agreement with a particular matter. Some of this can be seen in improved cross-Straits travel between Taiwan and China, as well as between the United States and some of its traditional adversaries (Kim *et al.* 2007).

In the past, several host countries severely restricted the types of tourists who could enter. For example, North Korea only recently began issuing special visas for US citizens, who until now were not permitted to enter. Similarly, Albania did not allow entry to US citizens until the 1990s, and for the nationals who could enter (e.g., British, European, Australian, etc.), stringent restrictions were placed on them regarding hair length, dress codes, and behavior (Hall 1984, 1992). Likewise, in many of the former state socialist countries of Eastern Europe and Asia, tourists could only travel with a guide, and the places they were allowed to visit or photograph were limited (Hall 1995). North Korea still has this same policy, and is one of the last countries on earth to conduct tourism in this manner (Kim *et al.* 2007).

A fifth use of tourism is to foster patriotism for the state among its citizenry (McLean 1998). This use has a great deal to do with heritage because, in most cases, a country's past, often its military or pioneer past, becomes a focal point for domestic tourism and education to develop nationalism. Domestic tourism to sites associated with national heroes, strategic battlefields, or other locations deemed important to the development of the state is often loaded information that endows these places with "national soul and memory for a unique purpose" (Timothy and Boyd 2003: 270). Thus, heritage places, artifacts, and tourism are employed as media that uphold national identity, legitimize governments in power, and reaffirm national ideologies (Fournier-Garcia and Miranda-Flores 1992; Hall 2002; Richter 1989; Ruiz 2002; Williams and Baláž 2001). This has also become a critical tool in idealizing the homeland among ethnic diasporic groups and in building support for the homeland abroad (Scheyvens 2007; Timothy 2007b). Such efforts are especially important following revolutions or civil wars in re-establishing national identity (Ruiz 2002).

Heritage and tourism are also used as an instrument for propagandizing foreign tourists. Places and their interpretations are thus geared to discredit negative events from the past, while extolling the virtues of the past and present, including governing ideologies. Most of the communist countries of Eastern Europe and Asia practiced this approach (Arefyev and Mieczkowski 1991; DeWald 2008; Murakami 2008). Under this guise, tourists are typically taken to factories, schools, historic monuments, and nationalist memorials that reinforce the ideals of the people in power and demonstrate the successes of a controversial political dogma. In Myanmar, government museums use the heritage of opium production to emphasize the evils of narcotics trafficking, demonstrating the decency of the regime in working to eliminate poppy production.

Finally, political manipulation is sometimes done to achieve the desired goal of erasing certain pasts from the heritage landscape of a country. Often governments will de-emphasize shameful, embarrassing, or revealing elements of history in order to control the story told to domestic and foreign tourists. Some administrations are loath to endorse former regimes that were responsible for certain less desirable heritages (Long 2002). Likewise, because place names are crucial in developing community collective memory (de Koninck 2003), some administrations will even change place names (e.g., villages and streets) to suggest, albeit subtly, that a certain heritage has never been. This "societal amnesia" (Timothy and Boyd 2003, 2006) may also occur unintentionally, but not including it in modern history is an intentional act. Among the most commonly cited examples are the indigenous pasts of Australia, the United States, and South Africa, where native cultures and heritages were suppressed by the white majorities for many years, and in places such as Malaysia, where the Chinese minority has been downplayed in its role in the country's development (Worden 2001). Slavery in the United States, the Caribbean, and Great Britain has also faced a similar challenge of being unheard until relatively recently, as social consciousness now demands that more holistic and accurate heritages are researched, displayed, and interpreted for tourists and other consumers (Dann and Seaton 2001; Teye and Timothy 2004; Timothy and Teye 2004). In Timothy's (2007b: xiii) words, "Unfortunately and predictably, most victims of societal amnesia have been ethnic and racial minorities, women and other 'marginal' peoples, and this has resulted in their lives and struggles being hidden from public view." This is a common problem throughout the world as leaders in power or colonial rulers suppressed the cultures and heritages of indigenous minority and majority populations. However, there is also a concomitant movement today to appreciate the heritage of everyday life, the struggles of the peasant past, and the mundane landscapes not associated with the wealthy elites (Grainge 1999; Timothy and Boyd 2006).

Power and empowerment

Endemic to the developing world is a tradition of centralized power, wherein grassroots planning and development and participatory governance have not been the normative practice (Aagesen 2000; Hall 2003; McNeely and Pitt 1985; Ormsby 1996; Timothy 1999, 2002; Wells and Brandon 1992). In most cases, development initiatives commence from central administration and are imposed on communities. This is neither healthy nor desirable. Modern conceptualizations of sustainable development, including tourism, argue that development initiatives should originate from the communities that will be most affected by them. Thus, control of the resources involved and the instigation of new tourism development initiatives need to rest in the hands of the destination communities themselves.

Empowerment, something often lacking in developing regions, indicates devolution of power from central authorities to individuals or communities. This practice, while relatively novel in much of the world, is assumed to support several principles of sustainable tourism development, including preserving ecological and cultural integrity, harmony, equity, and holistic growth (Davidson and Maitland 1999; Scheyvens 2002; Timothy and Tosun 2003). In the deepest sense, empowerment entails more than higher order governments simply allowing local communities and lower order administrators to be involved in the planning process or to benefit from tourism development. Rather, it indicates ownership of development programs and problems, including the consequences of wrong choices and mismanagement.

In response to this realization, Timothy (2007a) determined four degrees of empowerment that reflect destination communities in the developing world.

The first is imposed development, which can hardly be considered empowerment at all because it involves decisions taken at higher levels being imposed at the very grassroots level without consultation or other forms of public input. Although it does not resemble empowerment, it might nonetheless be the very initial phase on a larger trajectory toward full empowerment and control. Second is tokenistic involvement, wherein central planning authorities are beginning nascent efforts in involving proletariat participation in decision-making by asking for opinions or recommendations in various public forums. This sometimes happens as part of protocol or in response to global pressure to decentralize power rather than via any meaningful desire to involve destination residents. While this is a good beginning in participatory planning, it is far from creating empowered communities and citizens. The third level of empowerment is meaningful participation. This entails development agencies involving local residents and business people in decision-making processes and allowing them to benefit socio-economically from tourism. Development at this level is still controlled by governments or development agencies, but in their role they attempt to achieve meaningful dialog and discover solutions to real-life problems. Despite the positive step in the right direction, control still rests with outsiders. It resembles a grade of empowerment, but it is not enough only to seek opinions, as having a voice does not guarantee equality or empowerment. Lincoln *et al.* (2002: 285) suggest, "Individuals or groups that do not perceive that real power has been delegated are not empowered. They may hear the words, but when they see that the behavior is not consistent with the words, they rarely believe that empowerment has occurred."

Full empowerment signifies that community members initiate their own goals and programs for development. Of course, they may still rely on encouragement or assistance from non-governmental organizations (NGOs) and government leaders, but this is meant to facilitate rather than to lead (Logan and Moseley 2002). Empowered communities take responsibility for their own tourism development and take pride in the ownership of problems and solutions. Thus, development comes within reach and becomes more sustainable.

Just as various degrees or levels of empowerment exist, observers have also identified different types of empowerment. First is political empowerment. This occurs when communities and their residents have a voice in planning and policy-related decision-making. Political empowerment exists when development problem ownership and the benefits of tourism are located squarely in the hands of the destination community. This is especially important for traditionally marginalized peoples, such as indigenous groups, women, and racial minorities. As people gain power to determine their own futures, they become politically more empowered (Laverack and Wallerstein 2001; Scheyvens 1999; Timothy 1999).

Social empowerment, the second form, has the potential to enhance the destination community's solidarity and equilibrium (Cole 2006; Scheyvens 1999). Societal and individual self-esteem can be enhanced, and individual empowerment surrenders to a sense of group identity, or that a single person is part of a larger system of interdependent parts (Cousins and Kepe 2004; Lyons *et al.* 2001; Wilson 1996). Social empowerment is particularly important for societies in which social hierarchy exists based on caste or ethnicity. Indigenous knowledge and environmental practices are an important part of this and are now seen as a crucial element in development and policy-making. Today, native wisdom is believed to be an important tool for finding solutions to less-sustainable uses of resources (Briassoulis 2002; Khan 1997).

Psychological empowerment is similar to social empowerment in that community confidence grows when indigenous knowledge is utilized. This typically results in communities taking more pride in their traditions and an increased willingness to share their culture with outsiders (Scheyvens 1999). Psychological empowerment emboldens a sense of ownership and embeddedness, and it can help communities appreciate and respect their own heritage more profoundly (Timothy 2007a).

The final form of empowerment is economic in nature and is evident as tourism begins to produce true economic benefits for destination residents (Scheyvens 1999). This is one of the most important types, as it begins to allow the poor and other underprivileged people to benefit from tourism rather than only bear the costs of its growth and development. Jobs are created and income is received and shared. Likewise, increased income results in a number of other reciprocal benefits, such as improved education and health care facilities, better public services, and other opportunities to improve people's standard of living. While economic empowerment is an important and necessary outcome of all types of tourism, it is the pivotal basis for the current movement toward 'pro-poor' tourism and efforts toward poverty alleviation through tourism (Carbone 2005; Neto 2003; Renard 2001).

One developing world example of success in community empowerment in heritage and nature-based tourism is the Toledo Ecotourism Association (TEA), which was developed by Garifunas and indigenous Mayans in southern Belize to provide adventurous tourists with an alternative destination for heritage and accommodations. Members of the association decided to develop culture- and nature-based tourism in several villages. Certain cultural traditions, including hunting methods, food preparation and cuisine, village life, and ceremonies, were selected as the most appropriate heritage elements to be demonstrated to tourists. Small-scale accommodations were patterned after traditional homes were built by community members. The TEA is still functioning with the primary goal of widespread and equal economic benefits to all member villages and villagers. The program was initiated at the grassroots level and formed without interference from the national government, and is generally seen as an example of village-level sustainable development and empowered communities working together for the common good (Timothy and White 1999). Hatton (1999) also provides a number of burgeoning heritage tourism examples of meaningful participation and empowerment at the community level in China, the Philippines, Thailand, and Malaysia.

According to Xiao'an *et al.* (2003), the more empowered a community becomes, the more likely it should begin to demonstrate local control of resources and tourism. This is even the case at World Heritage Sites (WHS), where central governments might bear the primary burden and costs of maintaining and conserving the sites themselves, but it should be local authorities, private business people, and community residents who are in charge of developing tourism around the WHS.

Political instability

Political instability manifests in a variety of ways. Among the most commonly noted in the tourism literature are war, coup d'etat, and terrorism, although several others also exist, including natural disasters and their political ramifications, crime, corruption, minor disputes, and changing government regimes. Wars and other political upheavals have been examined in the context of tourism and have been shown to affect tourism negatively in a variety of ways, including reducing arrivals and invoking negative perceptions of the place in overseas markets (Sönmez 1998; Teye 1988). These events are also extremely destructive to historic places and heritage artifacts, especially when these are intentionally targeted for destruction by opposing forces (Bevan 2006), as was the case in the historic city of Dubrovnik during the 1990s Yugoslav wars (Oberreit 1996) and the now well-known case of Bamyan, Afghanistan—a tragic example where two large, sixth-century statues of Buddha were destroyed by the totalitarian Taliban rulers of that country in 2001 (Ashworth and van der Aa 2002).

During times of crisis, the status of heritage places often remains in question. Not only do historic assets suffer as targets, however, they also suffer from being considered dispensable items in the face of looming war or other conflict (Spennemann 1999). Thus, funds are necessarily diverted to other purposes (Phillips 1993) and often, when conflicts are over, new monuments are erected to replace the ones that existed before.

Although political conflict exists everywhere, the most notable examples as they pertain to heritage and tourism in the modern world have been identified in Southeast Asia, Eastern Europe, Africa, Latin America, and South Asia (Bhattarai *et al.* 2005; Gottesman 2004; Jetly 2003; Ospina 2006; Teye 1988). Aliçka (1997) noted that, in the wake of political changes in Albania from a state socialist system to a capitalist system, much damage was done to that country's heritage. Apparently millions of dollars of damage was done via looting and wanton destruction to cultural properties, such as municipal buildings, theaters, museums, and historic books and libraries. Pol Pot's Khmer Rouge regime was instrumental in eradicating not only the educated population, "but also all traces of national culture and history" (Masello 2001: 97; Peters 1994; Turnbull 1998). The ongoing battles in Afghanistan since the 1980s have destroyed much of that country's heritage landscape and historic urban centers (Dahlberg 1996), and the demolition of the Babari Masjid in India noted earlier resulted in a national crisis as Muslims and Hindus continued to retaliate against each other (Leech 2004). This conflict spread across the border into Pakistan, where Muslims retaliated by destroying Hindu temples there (Adil 1992).

Conclusion

The aim of this chapter was to discuss the political aspects of heritage and heritage tourism. Heritage and politics are intrinsically intertwined; it is almost impossible to exclude politics in discussions of heritage. The relationship between heritage and politics, however, is far more complex in developing countries than in more affluent parts of the globe, as conflicts, political instability, and, in many cases, centralized policies compound these relationships even further. Contestation is universal and intrinsic to the nature of heritage because of discordance and disagreement on how heritage is interpreted and produced by different groups of people (Tunbridge and Ashworth 1996). This chapter discussed three types of heritage contestation (Olsen and Timothy 2002). First, heritage contestation occurs when different groups claim the same heritage. Second, various divisions within a single group interpret and use the same heritage differently. The third type of contestation occurs when there is a parallel past within the same place.

Heritage is also intentionally used, manipulated, and sometimes abused by governments for the purpose of exercising power to control people and their places. Travel warnings, travel bans and embargos, and promises of increased arrivals from origin countries for compliance are common political uses of heritage and tourism frequently implemented by developed countries over developing countries. As of September 25, 2008, twenty-eight countries, all of them developing countries, are listed on the US Travel Warnings list (US Department of State 2008). Most of these travel warnings are political in nature. In addition to the tourists' origin countries, host countries in the developing world sometimes restrict certain types of tourists and prevent visits to certain regions for political purposes.

Tourism to heritage places is used to foster patriotism and national identity among the citizens of a country (McLean 1998), or to disregard negative events among foreign tourists through "societal amnesia" (Arefyev and Mieczkowski 1991; DeWald 2008; Murakami 2008). The common form of societal amnesia includes hiding and erasing shameful or embarrassing histories, or undesirable heritages through "historical consciousness" (Macdonald 2006). Although many examples of societal amnesia can be seen in the US, UK, Australia, South Africa, and many other developed countries where indigenous people were at some point in time oppressed by the white majority or white ruling minority, this is no less important in the developing world, where colonial powers suppressed the cultures and patrimony of the natives. Even today, many ethnic minorities have been marginalized and are coerced to adapt to the majority culture by official government policies in many developing nations.

Despite some political developments, many less-affluent countries still suffer from centralized and powerful elites, where administrations impose policies and plans without grassroots involvement. This has direct implications on what heritage should be preserved and managed, and for whom. True empowerment, a chief principle of sustainable development, results in increased pride as communities possess ownership of development projects and heritage preservation initiatives. This chapter argues that, in order for heritage tourism to succeed, community empowerment must exist in political, social, psychological, and economic forms. The role of NGOs is usually effective in involving local people and empowering them through various development and tourism-related projects (Nyaupane *et al.* 2006). Further, the commitments of many governments, communities, and international agencies to heritage conservation can easily be jeopardized by political instability and conflict, which are, unfortunately, frequent occurrences in many developing regions.

Despite the complexity of the relationship between heritage and politics, an understanding of at least some of the issues can help policy-makers and managers conserve and manage heritage more sustainably in the developing world. Creating public awareness through education may be a starting point that can empower communities and help them value and preserve their own unique heritage resources.

References

Aagesen, D. (2000) Rights to land and resources in Argentina's Alerces National Park. *Bulletin of Latin American Research*, 19: 547–69.

Adams, K.M. (2003) The politics of heritage in Tana Toraja, Indonesia: interplaying the local and the global. *Indonesia and the Malay World*, 31: 91–107.

Adil, A. (1992) Lahore: frenzied crusade. *Newsline*, 4(6): 36.

Aliçka, Y. (1997) Albania's threatened heritage. *The UNESCO Courier*, July/August: 78–79.

Arefyev, V. and Mieczkowski, Z. (1991) International tourism in the Soviet Union in the era of glasnost and perestroyka. *Journal of Travel Research*, 29(4): 2–6.

Ashworth, G.J. and van der Aa, B.J.M. (2002) Bamyan: whose heritage was it and what should we do about it? *Current Issues in Tourism*, 5(5): 447–57.

Bevan, R. (2006) *The Destruction of Memory: Architecture at War*. London: Reaktion.

Bhattarai, K., Conway, D. and Shrestha, N. (2005) Tourism, terrorism and turmoil in Nepal. *Annals of Tourism Research*, 32: 669–88.

Briassoulis, H. (2002) Sustainable tourism and the question of the commons. *Annals of Tourism Research*, 29(4): 1065–85.

Carbone, M. (2005) Sustainable tourism in developing countries: poverty alleviation, participatory planning, and ethical issues. *The European Journal of Development Research*, 17(3): 559–65.

Cartier, C. (1996) Conserving the built environment and generating heritage tourism in Peninsular Malaysia. *Tourism Recreation Research*, 21(1): 45–53.

Cole, S. (2006) Information and empowerment: the keys to achieving sustainable tourism. *Journal of Sustainable Tourism*, 14(6): 629–44.

Cousins, B. and Kepe, T. (2004) Decentralisation when land and resource rights are deeply contested: a case study of the Mkambati eco-tourism project on the wild coast of South Africa. *The European Journal of Development Research*, 16(1): 41–54.

Dahlberg, J.T. (1996) A rich heritage devastated. *Contours*, 7(8): 27–28.

Dann, G. and Seaton, A.V. (eds) (2001) *Slavery, Contested Heritage, and Thanatourism*. New York: Haworth.

Davidson, R. and Maitland, R. (1999) Planning for tourism in towns and cities. In C. H. Greed (ed.), *Social Town Planning*, pp. 208–20. London: Routledge.

de Koninck, R. (2003) Wessex Estate: recollections of British military and imperial history in the heart of Singapore. *Asian Journal of Social Science*, 31(2): 435–51.

DeWald, E. (2008) The development of tourism in French Colonial Vietnam, 1918–40. In J. Cochrane (ed.), *Asian Tourism: Growth and Change*, pp. 221–32. Amsterdam: Elsevier.

Emmett, C.F. (2000) Sharing sacred space in the Holy Land. In A.B. Murphy and D. L. Johnson (eds), *Cultural Encounters with the Environment: Enduring and Evolving Geographic Themes*, pp. 261–83. Lanham, MD: Rowman and Littlefield.

Fournier-Garcia, P. and Miranda-Flores, F.A. (1992) Historic sites archaeology in Mexico. *Historical Archaeology*, 26(1): 75–83.

Gottesman, E. (2004) *Cambodia after the Khmer Rouge: Inside the Politics of National Building*. New Haven, CT: Yale University Press.

Graham, B., Ashworth, G.J. and Tunbridge, J.E. (2000) *A Geography of Heritage: Power, Culture and Economy*. London: Arnold.

Grainge, P. (1999) Reclaiming heritage: colourization, culture wars and the politics of nostalgia. *Cultural Studies*, 13(4): 621–38.

Hall, C.M. (2003) Politics and place: an analysis of power in tourism communities. In S. Singh, D.J. Timothy, and R.K. Dowling (eds), *Tourism in Destination Communities*, pp. 99–114. Wallingford: CABI.

Hall, D. (1984) Foreign tourism under socialism: the Albanian "Stalinist" model. *Annals of Tourism Research*, 11: 539–55.

—— (1992) Albania's changing tourism environment. *Journal of Cultural Geography*, 12(2): 35–44.

—— (1995) Tourism change in Central and Eastern Europe. In A. Montanari and A. M. Williams (eds), *European Tourism: Regions, Spaces and Restructuring*, pp. 221–44. Chichester: Wiley.

—— (2002) Brand development, tourism and national identity: the re-imaging of former Yugoslavia. *The Journal of Brand Management*, 9(4): 323–34.

Hancock, M. (2002) Subjects of heritage in urban southern India. *Environment and Planning D: Society and Space*, 20: 693–717.

Hatton, M.J. (1999) *Community-based Tourism in the Asia-Pacific*. Ottawa: Canadian International Development Agency.

Henderson, J.C. (2003) Terrorism and tourism: managing the consequences of the Bali bombings. *Journal of Travel and Tourism Marketing*, 15(1): 41–58.

Henthorne, T.L. and Miller, M.M. (2003) Cuban tourism in the Caribbean context: a regional impact assessment. *Journal of Travel Research*, 42(1): 84–93.

Jayawardena, C. (2003) Revolution to revolution: why is tourism booming in Cuba? *International Journal of Contemporary Hospitality Management*, 15(1): 52–58.

Jetly, R. (2003) Conflict management strategies in ASEAN: perspectives for SAARC. *The Pacific Review*, 16(1): 53–76.

Kaneko, K. (1994) The safeguarding and promotion of Vietnamese ethno-forms: protecting the intangible cultural heritage of minority groups in Vietnam. *Vietnamese Studies*, 112: 44–50.

Keita, L. (1988) Africa's triple heritage: unique or universal? *Presence Africaine*, 143: 91–100.

Khan, M.M. (1997) Tourism development and dependency theory: mass tourism vs. ecotourism. *Annals of Tourism Research*, 24(4): 988–91.

Kim, S.S., Timothy, D.J. and Han, H.C. (2007) Tourism and political ideologies: a case of tourism in North Korea. *Tourism Management*, 28: 1031–43.

Laverack, G. and Wallerstein, N. (2001) Measuring community empowerment: a fresh look at organizational domains. *Health Promotion International*, 16(2): 179–85.

Leech, K. (2004) The ravage of India. *Blueprint*, 218: 86–91.

Lincoln, N.D., Travers, C., Ackers, P. and Wilkinson, A. (2002) The meaning of empowerment: the interdisciplinary etymology of a new management concept. *International Journal of Management Review*, 4(3): 271–90.

Logan, B.I. and Moseley, W.G. (2002) The political ecology of poverty alleviation in Zimbabwe's Communal Areas Management Programme for Indigenous Resources (CAMPFIRE). *Geoforum*, 33(1): 1–14.

Long, C. (2002) A history of urban planning policy and heritage protection in Vientiane, Laos. *International Development Planning Review*, 24(2): 127–44.

Löwenheim, O. (2007) The responsibility to responsibilize: foreign offices and the issuing of travel warnings. *International Political Sociology*, 1(3): 203–21.

Lowenthal, D. (1997) *The Heritage Crusade and the Spoils of History*. New York: Viking.

Lyons, M., Smuts, C. and Stephens, A. (2001) Participation, empowerment and sustainability: (how) do the links work? *Urban Studies*, 38: 1233–51.

Macdonald, S. (2006) Undesirable heritage: Fascist material culture and historical consciousness in Nuremberg. *International Journal of Heritage Studies*, 12 (1), 9–28.

McLean, F. (1998) Museums and the construction of national identity: a review. *International Journal of Heritage Studies*, 3(4): 244–52.

McNeely, J.A. and Pitt, D. (eds) (1985) *Culture and Conservation: The Human Dimension in Environmental Planning*. New York: Croom Helm.

Masello, D. (2001) Replanting the killing fields. *Art and Antiques*, 24(9): 96–99.

Mowry, D. (1999) Lifting the embargo against Cuba using Vietnam as a model: a policy paper for modernity. *Brooklyn Journal of International Law*, 25: 229–62.

Murakami, D. (2008) Tourism development and propaganda in contemporary Lhasa, Tibet Autonomous Region (TAR), China. In J. Cochrane (ed.), *Asian Tourism: Growth and Change*, pp. 55–68. Amsterdam: Elsevier.

Neto, F. (2003) A new approach to sustainable tourism development: moving beyond environmental protection. *Natural Resources Forum*, 27(3): 212–22.

Nyaupane, G.P., Morais, D.B., Dowler, L. (2006). The role of community involvement and number/type of visitors on tourism impacts: a controlled comparison of Annapurna, Nepal and Northwest Yunnan, China. *Tourism Management*, 27(6): 1373–85.

Oberreit, J. (1996) Destruction and reconstruction: the case of Dubrovnik. In D. Hall and D. Danta (eds), *Reconstructing the Balkans: A Geography of the New Southeast Europe*, pp. 67–77. Chichester: Wiley.

Olsen, D.H. and Timothy, D.J. (2002) Contested religious heritage: differing views of Mormon heritage. *Tourism Recreation Research*, 27: 7–15.

Ormsby, A.A. (1996) An attitudinal survey of resident perceptions of conservation at Five Blues Lake National Park, Belize. *TRI News*, 15(1): 13–15.

Ospina, G.A. (2006) War and ecotourism in the national parks of Colombia: some reflections on the public risk and adventure. *International Journal of Tourism Research*, 8: 241–46.

Peters, H.A. (1994) Cultural heritage at the National Museum of Cambodia. *Oriental Art*, 40(1): 20–27.

Phillips, D. (1993) Planning in the third world: a conservation project in Vietnam. *Australian Parks and Recreation*, 29(3): 14–17.

Renard, Y. (2001) *Practical Strategies for Pro-poor Tourism: a Case Study of the St. Lucia Heritage Tourism Programme*. Greenwich: Center for Responsible Tourism.

Richter, L. (1989) *The Politics of Tourism in Asia*. Honolulu, HI: University of Hawaii Press.

Ruiz, A. (2002) "La India Bonita": national beauty in revolutionary Mexico. *Cultural Dynamics*, 14(3): 283–301.

Scheyvens, R. (1999) Ecotourism and the empowerment of local communities. *Tourism Management*, 20: 245–49.

—— (2002) *Tourism for Development: Empowering Communities*. Harlow: Prentice Hall.

—— (2007) Poor cousins no more: valuing the development potential of domestic and diaspora tourism. *Progress in Development Studies*, 7(4): 307–25.

Sönmez, S. (1998) Tourism, terrorism and political instability. *Annals of Tourism Research*, 25: 416–56.

Spennemann, D.H.R. (1999) Cultural heritage conservation during emergency management: luxury or necessity? *International Journal of Public Administration*, 22 (5); 745–804.

Teye, V.B. (1988) Coup d'etat and African tourism: a study of Ghana. *Annals of Tourism Research*, 15: 329–56.

Teye, V.B. and Timothy, D.J. (2004) The varied colors of slave heritage in West Africa: white American stakeholders. *Space and Culture*, 7(2): 145–55.

Thapa, B. (2003) Tourism in Nepal: Shangri-La's troubled times. *Journal of Travel and Tourism Marketing*, 15(2/3): 117–38.

Thinh, N.D. (1999) Luăt tuc—a precious cultural heritage of ethnicities in the central highlands of Vietnam. *Vietnam Social Sciences*, 2: 23–29.

Timothy, D.J. (1999) Participatory planning: a view of tourism in Indonesia. *Annals of Tourism Research*, 26(2): 371–91.

—— (2002) Tourism and community development issues. In R. Sharpley and D.J. Telfer (eds), *Tourism and Development: Concepts and Issues*, pp. 149–64. Clevedon, UK: Channel View Publications.

—— (2007a) Empowerment and stakeholder participation in tourism destination communities. In A. Church and T. Coles (eds), *Tourism, Power and Space*, pp. 199–216. London: Routledge.

—— (2007b) Introduction. In D.J. Timothy (ed.), *The Political Nature of Cultural Heritage and Tourism*, pp. ix–xviii. Aldershot: Ashgate.

Timothy, D.J. and Boyd, S.W. (2003) *Heritage Tourism*. Harlow: Prentice Hall.

—— (2006) Heritage tourism in the 21st century: valued traditions and new perspectives. *Journal of Heritage Tourism*, 1(1): 1–16.

Timothy, D.J. and Emmett, C.F. (forthcoming) Jerusalem, tourism and the politics of heritage. In M. Adelman and M. Ellman (eds), *Jerusalem across Disciplines*, Bloomington, IN: Indiana University Press.

Timothy, D.J. and Teye, V.B. (2004) American children of the African diaspora: journeys to the motherland. In T. Coles and D.J. Timothy (eds), *Tourism, Diasporas and Space*, pp. 111–23. London and New York: Routledge.

Timothy, D.J. and Tosun, C. (2003) Appropriate planning for tourism in destination communities: participation, incremental growth and collaboration. In S. Singh, D.J. Timothy and R.K. Dowling (eds), *Tourism in Destination Communities*, pp. 181–204. Wallingford: CABI.

Timothy, D.J. and White, K. (1999) Community-based ecotourism development on the periphery of Belize. *Current Issues in Tourism*, 2(2/3): 226–42.

Tunbridge, J.E. and Ashworth, G.J. (1996) *Dissonant Heritage: The Management of the Past as Resource in Conflict*. Chichester: Wiley.

Turnbull, R. (1998) A fragile dance. *Far Eastern Economic Review*, 161(52): 36–38.

US Department of State (2008). Current travel warnings. Available from http://travel.state.gov/travel/cis_pa_tw/tw/tw_1764.html (accessed September 25, 2008).

Weinmann, L. (2004) Washington's irrational Cuba policy. *World Policy Journal*, 21: 22–31.

Wells, M. and Brandon, K. (1992) *People and Parks: Linking Protected Area Management with Local Communities*. Washington, DC: World Bank.

Williams, A.M. and Baláž, V. (2001) From collective provision to commodification of tourism? *Annals of Tourism Research*, 28(1): 27–49.

Wilson, P. (1996) Empowerment: community economic development from the inside out. *Urban Studies*, 33: 617–30.

Worden, N. (2001) "Where it all began": the representation of Malaysian heritage in Melaka. *International Journal of Heritage Studies*, 7(3): 199–218.

Xiao'an, W., Qun, D. and Decheng, P. (2003) On sustainable tourism of China's world-class cultural and natural heritage sites. *Social Sciences in China*, 24(1): 160–68.

4 Heritage tourism and its impacts

Introduction

Many developing countries have focused on tourism to promote economic growth in economically depressed regions and to enhance the socio-economic well-being of their people (Nyaupane *et al.* 2006). As a result, tourism has emerged as a crucial contributor to 70 percent of the world's poorest countries (Lipman 2008). Although these countries are economically poor, they are characterized as culturally rich destinations with many ethnic groups, traditions, religions, and languages, which constitute a major attraction for millions of tourists. However, when people's private community and sacred spaces are open to masses of tourists, these places may experience negative consequences. Most of these consequences or impacts associated with heritage-based tourism are the same in the developed and developing worlds. However, they tend to be more pronounced in the less-developed parts of the world. Therefore, most of the information in this chapter could be applied to any world context, although less-developed regions will be used to illustrate many of the points being made. This chapter first examines briefly the impacts of tourism in general and then places them into a heritage context.

The impacts of tourism

This chapter highlights the negative and positive impacts associated with cultural/heritage tourism, which have traditionally been divided into physical or environmental, socio-cultural, and economic (Mathieson and Wall 1982). Most tourism textbooks and studies undertaken on the impacts of the industry utilize this same framework, as all elements, tangible and intangible, of the "world around us" fit somewhere within this tripartite environment.

The physical environment refers to most elements of the physical world. Built structures, rocks and bedrock, soil, vegetation, water, and air comprise this segment of the environment. The socio-cultural realm is typically synonymous with intangibles, such as music, dance, traditions, religious beliefs, education, foodways, and social mores, but it also includes cultural artifacts, such as artworks, handicrafts, apparel, and food products. The economic environment is the most intangible of the three and includes economic systems, fiscal policies, taxation, employment, and funding. Tourism affects all three of these environmental divisions in both positive and negative ways (Flemming and Toepper 1990; Mason 2003; Mathieson and Wall 1982; Wagner 1996; Wall and Mathieson 2006).

Socio-culturally, tourism is seen as a force that damages cultures, generates prostitution, drug addiction, gambling obsessions, and alcoholism. However, it has also been shown to function as a medium by which cultures can be protected and lost social celebrations resurrected. In physical terms, tourism is blamed for a great deal of wear damage to rocks and built heritage, vandalism, air and water pollution, fires, and soil compaction, but it is also a tool for funding conservation. Tourism even has negative economic implications, such as inflation, overdependence, monetary leakage, a tendency to widen the gap between the haves and the have-nots, and low-wage earnings, although most economic impacts are seen in a positive light: increased regional income, employment generation, tax revenues, and stimulation of entrepreneurialism. The next section examines some of these issues and others related more specifically to the effects of cultural and heritage tourism.

The impacts of heritage tourism

The same set of impacts noted above can be narrowed and examined in the context of heritage places. In the realms of physical and socio-cultural environments, there are several unique impacts that apply well to cultural heritage tourism. Each of these is examined below.

Physical impacts

Excessive or careless visitor use of historic artifacts and ancient monuments causes serious damage, as many research studies have noted (Austin 2002; Fyall and Garrod 1998; Merhav and Killebrew 2003; Timothy and Boyd 2003). Unfortunately, large masses of tourists and their oftentimes injudicious behavior have been shown to deteriorate the very objects and places that attracted them in the first place. Wall (1989: 10) noted, "it is a paradox that participants in [tourism] are drawn to attractive environments, whether natural or built, but that their mere presence is likely to result in the modification of those environments." Through this process, tourists make the attraction less attractive for themselves. Timothy (1994) identified several specific impacts associated directly with visitor use at heritage sites. These are described in detail below.

One of the most serious effects of tourism on the physical heritage environment is wear and tear. The gravity of this issue cannot be overstated because, unlike many components of natural environments, historic human environments are non-renewable resources that cannot be regenerated organically. The deterioration of the built environment occurs in a number of ways, although the most direct occurs when visitors touch, climb on, or rub historic structures and artifacts.

Until the late 1970s, tourists were permitted to walk in and around the stones at Stonehenge. As a result of some 2,000 people per hour touching, leaning, and climbing on the structure during high season and many others during off seasons, the ruins began to suffer visible degradation. The stones were being worn smooth and the earthworks surrounding them were eroding badly, resulting in the stones beginning to lean. In 1978, protective measures were enacted by the Department of the Environment to restrict tourist access to the site (Bainbridge 1979). Although not as well documented in the less-developed world, many examples also exist there. Steps similar to those taken at Stonehenge have been taken in Egypt to address similar problems at the Pyramids. Tourists' physical contact has caused so much damage to the Egyptian Pyramids that site managers have had to restrict visitor access to save the relics from further deterioration. Similarly, in the Valley of the Kings near Luxor, Egypt, once the tombs were opened to tourists, millions visited, resulting in altered air conditions inside the tombs and rapid deterioration of wall paintings and carvings. The same happened at the tomb of Nefertari, causing Egyptian authorities to close the tomb to prevent further corrosion (Hang and Kong 2001).

Machu Picchu, Peru's most recognized tourist attraction and one of the best known heritage sites in the world, has come under threat since the 1970s and 1980s. Its increasing popularity among tourists—some estimates suggest 2,200 people per day—has resulted in notable wear and tear and garbage strewn about. Nearby villages that depend on the site for their tourism industries are also overrun with garbage and other tourism-related pollution, causing additional damage to Machu Picchu and its surroundings (*Economist* 2001).

This is a rampant problem at world-renowned places such as Borobudur, Angkor Wat, Prambanan, the Great Wall of China, and many others (Timothy 1999). The difficulty is finding a suitable balance between visitor use and conservation, which is difficult, because so many tourists desire to interact with the artifacts in a variety of ways. While many site managers have devised creative ways of limiting contact between visitors and artifacts, it still is not an exact science, and not all methods work at all locations.

In addition to direct structural damage, surrounding green spaces and landscaping are also affected, suffering considerable wear and tear as visitors veer from sidewalks and other prescribed pathways. This often results in compacted soils, where little will grow, and increased soil erosion (Mathieson and Wall 1982).

Litter is another problem associated with heritage places. Garbage is a major problem everywhere, although environmental regulations, staffing shortages, and budget restraints make clean-up more difficult in developing regions. Food containers, leftover food, cigarette butts, chewing gum, plastic bottles, aluminum soda cans, paper products, and even dirty diapers are all commonly present at heritage sites. Not only are these unsightly, reducing the esthetic appeal of a place, but they can and do contribute to the material corrosion of delicate properties. Some litter, such as chewing gum, drinks, or other food products, can permanently stain stone or ceramic surfaces.

Ritual litter is common in and around religious places. In many religious destinations, local residents and traveling pilgrims burn candles, worship with flowers and other accoutrements, and even sacrifice animals. Although ritual litter is a problem in many religious spaces, it is especially notable at Hindu sacred sites. For instance, one of the major sources of pollution in the Ganges River at Varanasi, India, is ritual remains such as flowers and *prasad* (food), as well as ashes and the remains of cremated corpses. All of this combines to create one of the most polluted sacred heritage sites in the world (Alley 1998; Cumming 2003).

Air pollution contributes to the dilapidation of heritage sites as well, as exhaust from buses and cars causes chemical reactions in building stones and materials. In India, the physical composition of the Taj Mahal, one of the world's most beloved monuments, has suffered grossly from the pollution discharge associated with factories and other heavy industrial developments in Agra and Delhi (Gauri and Holdren 1981). Fortunately, the federal and state governments have reacted to the situation and removed many of the most inefficient and polluting refineries from the vicinity of the Taj (Kalas 2000). Now, however, the vicinity suffers from excess vehicle pollution, including almost countless numbers of taxis, cars, motorcoaches, and motorized rickshaws as tourism in India has grown in recent years. While it is unlikely that tourism-related vehicles will cause as much damage to the Taj as the factories and refineries did, the level of pollution in Agra is still a significant concern, especially given the already delicate condition of the monument.

Human beings have an innate desire to take mementos home from their travels (souvenir-hunting) and to leave marks or remnants of themselves in places visited. Frequently the result is vandalism, which is a salient concern for all heritage managers. It manifests in a variety of ways, including breaking pieces of buildings or statues, spray-painting over sculpted reliefs, carving names or slogans, or burning. All of these disastrous behaviors can be witnessed at heritage sites. At one ninth-century temple on the Dieng Plateau in Indonesia, "The Reeves" was witnessed by Timothy (1994) carved deeply into the original construction stones. The Reeves family clearly wanted to leave its mark in Indonesia and, in so doing, caused irreparable damage to a site that is considered to be of high scientific, historical, and esthetic value. Such problems are very difficult to resolve, as clean-up efforts can be more damaging than the vandalism, especially if the only solution is using corrosive chemicals or sand-blasting.

A related concern, which was addressed in much more detail in Chapter 2, is the idea of illicit trade in artifacts. Many tourists who visit ancient sites have an interest in acquiring relics associated with the site. Their making this known to local people can result in destination residents undertaking illegal digs and antiquities theft from protected sites to sell to foreign visitors. Thus, tourism helps drive the illicit trade in antiquities, thereby destroying and removing not only ambulatory artifacts, but in the process, careless digging and stealing often causes critical damage to the archeological sites themselves (Lafont 2004; Melwani 2000; Prott 1996).

While the issues noted so far in this section are negative, not all the impacts of tourism on the built environment are harmful. Tourism brings a considerable sum of money into countries, regions, and communities that can be utilized to help preserve heritage and culture. In fact, in most developing countries, cultural heritage and natural areas are highly dependent upon tourism income for their survival, owing to a lack of public and private funds (Blom 2000; Cochrane and Tapper 2006; Wilkie and Carpenter 2002).

Tourism is also partially responsible for effecting awareness among community members and site managers of the importance of conserving the built environment (Timothy 2000). Similarly, tourists are beginning to demand more sophisticated and in-depth interpretive methods, which has led to the development of more informative, accurate, and educative interpretation programs. This is important in teaching visitors about the importance of heritage sites for, according to many observers, interpretation is the best form of education in the context of heritage (Butler 1990; Cossons 1989; Millar 1989; Timothy and Boyd 2003). In addition, interpretation can alleviate some of the physical pressures of tourism at monuments and sites by directing visitors away from sensitive and crowded areas, and encouraging them to refrain from climbing and touching (Moscardo 1996).

Socio-cultural impacts

Not all impacts of culture- and heritage-based tourism are physical or environmental in nature. Other socio-cultural results also emerge when tourists arrive at the destination and interact with local populations. Although many of these have been noted over the years in all tourism contexts (Mathieson and Wall 1982), some of them have direct links with cultural heritage.

In contexts of heritage-based tourism, tension between destination residents and tourists/tourism is not uncommon, like that in other tourism situations. The development of historic places into tourism resources is one primary culprit. Many people in the developing world depend on archeological sites for their livelihoods. This was discussed in Chapter 2. In some cases, communities have grown within and around ancient ruins. When these resources are developed for tourism purposes, conflicts usually occur, as inhabitants see this as an encroachment on their private spaces, traditional homes, and cultures.

The most significant problem related to this and among the most deplorable social outcomes of tourism is the forced displacement of local populations. Heritage conservation and tourism have a major role to play in the forced relocation of indigenous and powerless populations. There are many accounts of forced population displacements in an effort to develop tourism or in the name of "conserving" natural or cultural heritage (Guha 1997; Kasim 2006; McLean and Stræde 2003; Meskell 2005; Mortensen 2006; Mowforth and Munt 1998; Pandey et al. 1995; Parnwell 1998; Singh and Singh 2004; Wang

2004; Wang and Wall 2005). Mowforth and Munt (1998) and Timothy (1999) note that this clearly reflects the distribution of power in the developing world—the powerfulness of the ruling class and the powerlessness of the masses. Rarely are displaced populations able to reconnect to their new lands nor are they able to attain a significant level of prosperity.

Villages near the Borobudur and Prambanan temple complexes in Indonesia were forcefully removed to make way for the development of these sites for tourism. In one instance, villagers were relocated to the far eastern end of Java, some 600 kilometers from their traditional lands. Community members who resisted were intimidated by the military, and the compensation divvied out by the state was inadequate for the lands and homes being surrendered under duress (Timothy 1994, 1999). In many cases, no compensations are offered. Flynn (1996, cited in Mowforth and Munt 1998) noted a similar situation in Guatemala, where 300 families were expelled from their homes in 1996 to make way for a tourist development project. Police reportedly burned their homes and arrested some of those who resisted. Regrettably, torture, bullying, defrauding, imprisonment, and even murder are often used to subdue resistance to forced migration carried out for the sake of tourism development and to clear heritage areas for conservation.

Conflict also runs rampant in the realm of religious tourism. Some of the most historic buildings and locations in the world today of significant tourist interest are markers of spiritual or religious significance. Examples include the multitudinous Buddhist and Hindu temples of South and Southeast Asia, the churches and cathedrals of Latin America, and the mosques and temples of North Africa and Southwest Asia. Visits to these sacred sites are another source of discontent in tourism-dependent communities. While on the surface it seems logical that religious-based heritage tourism would spawn peaceful or benevolent relations, the opposite is frequently the case (Timothy and Olsen 2006). Friction at religious sites occurs in a few different ways. First, destination residents view all outsiders (even devout tourists, or pilgrims) with a degree of contempt for the same reasons noted in other tourism contexts (i.e., mischievous behaviors, pollution, inflation, crowdedness, etc.). Second, local devotees and pilgrim tourists utilize sacred spaces and artifacts for spiritual purposes: worship, prayer, meditation, healing, reading, chanting, singing, and resting. However, these same sacred spaces and objects attract large numbers of non-believers as well. The boisterousness, flashing cameras, irreverence, immodest dress standards, and religious ignorance among non-pilgrim tourists are highly offensive to worshippers and detract from the spirit of the place.

Cultural change is also often cited as one of the most salient negative impacts of tourism (Brunt and Courtney 1999; Mansperger 1995; Mathieson and Wall 1982), although some scholars and their study populations have argued that not all cultural modifications are bad (Ashley *et al.* 2000; Chang 2002). While it is common knowledge today that tourism is only one force in a wide array of modernizing influences, there is common agreement among anthropologists and other cultural studies scholars that tourism, including heritage tourism, is partially responsible for destination societies losing cultural traditions or undergoing cultural modifications (Mathieson and Wall 1982; Smith 1989; Woods *et al.* 1994).

One of the most often-cited side-effects of tourism is cultural commodification, whereby culture becomes a product that is packaged and sold to tourists (Cohen 1988; Hughes-Freeland 1993; Medina 2003; Timothy and Boyd 2003). In more traditional societies, this issue is especially acute. In the process of commoditization, the spiritual meanings or customary values behind traditional celebrations, music, dances, and handicrafts are often lost as these cultural elements begin to be mass produced for tourist consumption. In the words of Tilley (1997: 81, cited in Kirtsoglou and Theodossopoulos 2004), it results in a setting and a performance divorced from most aspects of native culture and becomes "an empty vessel of tradition ... form without sentiment." Traditional art forms are altered to meet the needs of tourists, and they too lose their value and become mass produced, meaningless, and inauthentic tourist kitsch (Cohen 1992, 1993; Graburn 1984; Urry 2002). This realization led MacCannell (1973) to suggest early on that, as a way of protecting rituals, customs, and traditional institutions, societies will often "stage" or present some kind of superficial illustration or cultural snippets to tourists as a way of protecting the "backstage", or the real-life practices and mores of a given society. Thus, cultural performances for tourists are inauthentic but nonetheless satisfy tourists' need for culture.

This theory has been variously contested and applied to many different socio-cultural contexts. Something that tends to run through the thread of all the case studies in the literature is that indigenous people in the developed world often have the luxury of staging their culture (e.g., some Native American tribes), thereby sheltering much of it from the tourist gaze (Jenkins *et al.* 1996; Lujan 1998). In the less-developed world, on the other hand, such a luxury rarely exists, and cultures in their rawest forms become spectacles for outsiders to view, mimic, and "desecrate."

Cultural commodification frequently results in the loss of control over cultural resources as outside agents begin to capitalize on cultural elements that belong to others. Centralized, top-down planning and development traditions in many parts of the world have precluded grassroots-level empowerment and communities' decisions regarding what can and ought to be shown to tourists and what should be kept from them. Many elements of aboriginal and minority cultures, for instance, are adopted into the commercial development of the state. This typically occurs without the approval or authorization of the people whose heritage is being consumed and results in conflict over rights of ownership and fair trade in culture (Johnston 2003, 2006; Timothy and Prideaux 2008). In most parts of the world, ethnic minorities have lacked control over how their cultures are represented to tourists (Goudie *et al.* 1999).

This lack of true ownership of culture is one of the most often-cited frustrations among indigenous peoples and local community members in cultural tourism destinations (Johnston 2006; Timothy and White 1999), where culture as a tourism resource is controlled by outsiders. Most commentators today, in line with the principles of sustainable development, argue that the use of a society's culture should be done on that society's terms and according to what they deem to be appropriate use.

Another related frustration is that outsiders, including tourists, national-level tourism promoters, filmmakers, and researchers come to villages and traditional settlements, "take the culture away," and profit from it without giving back to the community, which in most cases is in dire economic circumstances. According to people on the island of Roatan, Honduras, tourists and other outsiders:

> come and take Garifuna culture without giving anything back ... They came and made a film. Here in the village and the rest of the island. They shot [staged] pictures of people killing one another and many other things ... The Garifuna took nothing for this film. Everyone comes and makes money with Garifuna culture.
>
> Kirtsoglou and Theodossopoulos (2004: 145)

In addition to the profiteering aspect of this misappropriation of culture, outside use and control can also lead to stereotypes and false perceptions. For example, the Garifuna people of Roatan, Honduras, oppose their culture being used symbolically by the Honduran government and filmmakers for the country as a whole primarily because their culture is usually misrepresented and, as one resident put it regarding outside promoters, "they are irresponsible and lie about peculiar beliefs, such as magic and superstition ... the Garifuna people do not want to be known for those matters" (Kirtsoglou and Theodossopoulos 2004: 144).

In common with the physical impacts of heritage tourism, not all socio-cultural effects are negative. Several commentators have observed tourism as a positive force in reviving lost or declining elements of culture (Kolås 2004; Rogers 2002; Smith 2003). In the case of the Maasai of East Africa, cultural tourism has been a catalyst for reviving traditional celebrations, music, and dance (Irandu 2004). In Vietnam, heritage tourism has been instrumental in preserving cultural festivals (Son 2004). Societal self-esteem and cultural pride are often increased through tourism as local people present their chosen ethnic elements to outsiders (Ashley *et al.* 2000; Wood 1993). Likewise, new high-quality art forms that reflect modern societies' mores and values have been introduced, which are different from the commoditized "tourist kitsch" noted earlier and is typically seen as something positive in the process of modernization, because it expands knowledge, creates innovation, and provides employment for promising artisans (Ivory 1999; Ng 2002).

Economic impacts

Most of the economic outcomes of cultural and heritage tourism are the same or similar to those of other forms of tourism. Perhaps worth noting here, however, is the fact that increased regional income and a broadened tax base through tourism have the potential to be instrumental in conserving, managing, and interpreting heritage sites. The most direct form of economic impact in heritage tourism is revenue regenerated through entrance fees. Experiences from many cities, such as Cairo, Tunis, Delhi, Galle, and Sana'a, show that most local and national governments and religious organizations cannot afford to conserve and improve their heritage sites (Steinberg 1996). Although entrance fees have been criticized from an equity perspective, fees to visit galleries, museums, historic buildings, ruins, and monuments are minimal and do not necessarily impact visitor numbers. Although entrance fees are a viable source of funding, many countries in the developing world do not have adequate policies directing entrance fees to heritage sites, as many of them are heavily used by locals. A few cities have adopted a more progressive policy in collecting revenue from tourists (Steinberg 1996). For example, the ancient city of Bhaktapur, Nepal, a living World Heritage Site, charges a US$10 entrance fee only for tourists, which is used for conserving and maintaining the site and its surroundings. Nepalese can enter without paying a fee. The revenue is also used to provide amenities and services to local residents.

Tourism activities in and around heritage sites stimulate the economies of neighboring communities through employment and private businesses. For developing countries, tourism is an especially important source of foreign currency. However, local residents often lack the skills and investment abilities needed to establish tourism-related businesses and often end up with low-investment businesses and low-paying jobs. Therefore, as discussed in Chapter 2, the public earnings associated with tourism in the developing world are still in short supply, as they are diverted to other priorities.

Conclusion

This chapter has discussed the impacts of heritage tourism, which have traditionally been grouped into three domains: physical, social, and economic. Physical impacts include erosion and corrosion of historic structures and artifacts when tourists and local residents overuse them. Many tourism impacts on the natural environment can recover through time as vegetation renews itself, whereas the impacts on the built environment are cumulative and permanent. This is an especially austere problem in countries and regions where protective measures and political will are lacking. In addition, many heritage sites suffer from tourist and resident litter, which in some cases is cleaned up regularly by street sweepers, but more often than not sits and accumulates.

In most cases, historic sites are crowded, and even small impacts by each individual visitor are compounded to become large problems. Illicit trade in antiquities and other artifacts is a common problem in developing countries, given the poverty associated with the public and the global demand for antiquities.

Unlike physical impacts, the social impacts of heritage tourism are hard to measure. Tourists often enter locals' sacred space to watch and photograph rituals. Local residents often become annoyed and feel that their sacred or personal space is being transgressed by non-believers or outsiders. This often results in conflict when the behavior of cultural tourists becomes intolerable to local worshippers and pilgrims. There is also a threat of over-commodification of culture, when destination populations perform rituals for tourists, which may in the process lose their original social or religious value (Greenwood 1982). This threat is even greater in less-developed regions owing to a lack of skills, resources, and awareness. In the name of heritage preservation and tourism development, many communities in developing countries are also forcefully displaced from their homes and villages. Unfortunately, this happens all too often and calls into question issues of equality, stability, sustainable development, and cultural property rights.

It is also important to note that not all impacts of heritage tourism are negative. Many sites have benefited greatly from heritage tourism. Tourism can be a source of cultural revival through societal self-esteem and pride. Economic benefits are the most important reasons behind the development and promotion of heritage tourism. It provides much needed economic incentives to communities and governments. Heritage tourism is especially prone to creating employment, providing infrastructure and public services for the community, helping fund the preservation and management of historic sites, and overall stimulating the local and national economy.

Despite this potential, a major challenge facing governments in the developing world is balancing the economic benefits of heritage tourism, or any tourism for that matter, against its negative implications. When governments emphasize only the short-term goal of maximizing profits without taking into consideration tourism's consequences, the result will almost always be insurmountable problems. Thousands of examples of this failure exist throughout the world. However, the good news is that more places are increasingly beginning to understand the value of sustainable tourism and sustainable development within the context of tourism (Butler 1999). Developing world governments are desperate for financial support from investors, aid agencies, and international organizations. As a result, they often undergo a resurgence of neo-colonialism in that they once again become dependent upon outside funding and control of their own national heritage.

References

Alley, K.D. (1998) Images of waste and purification on the banks of the Ganga. *City and Society*, 10(1): 167–82.

Ashley, C., Boyd, C. and Goodwin, H. (2000) Pro-poor tourism: putting poverty at the heart of the tourism agenda. *Overseas Development Institute Natural Resource Perspectives*, 51: 1–6.

Austin, N.K. (2002) Managing heritage attractions: marketing challenges at sensitive historical sites. *International Journal of Tourism Research*, 3(6): 447–57.

Bainbridge, S. (1979) *Restrictions at Stonehenge: The Reactions of Visitors to Limitations in Access.* London: HMSO.

Blom, A. (2000) The monetary impact of tourism on protected area management and the local economy in Dzanga-Sangha (Central African Republic). *Journal of Sustainable Tourism*, 8(3): 175–89.

Brunt, P. and Courtney, P. (1999) Host perceptions of sociocultural impacts. *Annals of Tourism Research*, 26: 493–515.

Butler, R.W. (1990) Tourism, heritage and sustainable development. In J.G. Nelson and S. Woodley (eds), *Heritage Conservation and Sustainable Development*, pp. 49–66. Waterloo: Heritage Resources Centre, University of Waterloo.

—— (1999) Sustainable tourism: a state-of-the-art review. Available from http://hdr.undp.org/en/media/HDR_20072008_EN_Chapter2.pdf, 1(1): 7–15.

Cabeza, A. (2001) Evaluating the environmental impact of development projects on the archaeological heritage of Chile. *Conservation and Management of Archaeological Sites*, 4(4): 245–47.

Chang, T.C. (2002) Heritage as a tourism commodity: traversing the tourist–local divide. *Singapore Journal of Tropical Geography*, 18(1): 46–68.

Cochrane, J. and Tapper, R. (2006) Tourism's contribution to World Heritage Site management. In A. Leask and A. Fyall (eds), *Managing World Heritage Sites*, pp. 97–109. Oxford: Butterworth Heinemann.

Cohen, E. (1988) Authenticity and commoditization in tourism. *Annals of Tourism Research*, 15: 371–86.

—— (1992) Tourist arts. *Progress in Tourism, Recreation and Hospitality Management*, 4: 3–32.

—— (1993) The heterogeneization of a tourist art. *Annals of Tourism Research*, 20: 138–63.

Cossons, N. (1989) Heritage tourism—trends and tribulations. *Tourism Management*, 10(3): 192–94.

Cumming, D. (2003) *The Ganges.* Strongsville, OH: Gareth Stevens.

Economist (2001) Road to ruin. *Economist*, July 21: 30.

Flemming, W.R. and Toepper, I. (1990) Economic impact studies: relating the positive and negative impacts to tourism development. *Journal of Travel Research*, 29(1): 35–42.

Flynn, M. (1997) Report on Guatemala. *Mesoamerica*, 15(8): 3–4.

Fyall, A. and Garrod, B. (1998) Heritage tourism: at what price? *Managing Leisure*, 3 (4): 213–28.

Gauri, K.L. and Holdren, G.C. (1981) Pollutant effects on stone monuments. *Environmental Science and Technology*, 15: 386–90.

Goudie, S.C., Khan, F. and Kilian, D. (1999) Transforming tourism: black empowerment, heritage and identity beyond apartheid. *South African Geographical Journal*, 81(1): 22–31.

Graburn, N.H.H. (1984) The evolution of tourist arts. *Annals of Tourism Research*, 11: 393–419.

Greenwood, D.J. (1982) Tourism to Greenland: renewed ethnicity. *Cultural Survival Quarterly*, 6(3): 26–28.

Guha, R. (1997) The authoritarian biologist and the arrogance of anti-humanism: wildlife conservation in the third world. *The Ecologist*, 27(1): 14–20.

Hang, L.K.P. and Kong, C. (2001) Heritage management and control: the case of Egypt. *Journal of Quality Assurance in Hospitality and Tourism*, 2(1/2): 105–17.

Hughes-Freeland, F. (1993) Packaging dreams: Javanese perceptions of tourism and performance. In M. Hitchcock, V.T. King and M.J.G. Parnwell (eds), *Tourism in Southeast Asia*, pp. 138–54. London: Routledge.

Irandu, E.M. (2004) The role of tourism in the conservation of cultural heritage in Kenya. *Asia Pacific Journal of Tourism Research*, 9(2): 133–50.

Ivory, C.S. (1999) Art, tourism and cultural revival in the Marqueses Islands. In R.B. Phillips and C.B. Steiner (eds), *Unpacking Culture: Art and Commodity in Colonial and Postcolonial Worlds*, pp. 316–34. Berkeley, CA: University of California Press.

Jenkins, L., Dongoske, K.E. and Ferguson, T.J. (1996) Managing Hopi sacred sites to protect religious freedom. *Cultural Survival Quarterly*, 21(1): 36–38.

Johnston, A.M. (2003) Self-determination: exercising indigenous rights in tourism. In S. Singh, D.J. Timothy and R.K. Dowling (eds), *Tourism in Destination Communities*, pp. 115–34. Wallingford: CABI.

—— (2006) *Is the Sacred for Sale? Tourism and Indigenous Peoples*. London: Earthscan.

Kalas, P.R. (2000) Environmental justice in India. *Asia-Pacific Journal on Human Rights and the Law*, 1: 97–116.

Kasim, A. (2006) The need for business environmental and social responsibility in the tourism industry. *International Journal of Hospitality and Tourism Administration*, 7 (1): 1–22.

Kirtsoglou, E. and Theodossopoulos, D. (2004) "They are taking our culture away": tourism and culture commodification in the Garifuna community of Roatan. *Critique of Anthropology*, 24(2): 135–57.

Kolås, Å. (2004) Tourism and the making of place in Shangri-La. *Tourism Geographies*, 6(3): 262–78.

Lafont, M. (2004) *Pillaging Cambodia: The Illicit Traffic in Khmer Art*. Jefferson, NC: McFarland and Co.

Lipman, G. (2008) Emerging tourism markets—the coming economic boom. Keynote speech delivered at the UK Tourism Society Annual Meeting, June 20. Available from http://www.tourismsociety.org/Conference%2008/Geoffrey%20Lipman.pdf (accessed January 5, 2009).

Lujan, C.C. (1998) A sociological view of tourism in an American Indian community: maintaining cultural integrity at Taos Pueblo. In A.A. Lew and G.A. Van Otten (eds), *Tourism and Gaming on American Indian Lands*, New York: Cognizant Communications.

MacCannell, D. (1973) Staged authenticity: arrangements of social space in tourist settings. *American Journal of Sociology*, 79: 589–603.

McLean, J. and Stræde, S. (2003) Conservation, relocation, and the paradigms of park and people management—a case study of Padampur Villages and the Royal Chitwan National Park, Nepal. *Society and Natural Resources*, 16: 509–26.

Mansperger, M.C. (1995) Tourism and cultural change in small-scale societies. *Human Organization*, 54(1): 87–94.

Mason, P. (2003) *Tourism Impacts, Planning and Management*. Oxford: Butterworth Heinemann.

Mathieson, A. and Wall, G. (1982) *Tourism: Economic, Physical and Social Impacts*. London: Longman.

Medina, L.K. (2003) Commoditizing culture: tourism and Maya identity. *Annals of Tourism Research*, 30: 353–68.

Melwani, L. (2000) Looking for the perfect Ganesha. *Art & Antiques*, 23(9): 68–73.

Merhav, R. and Killebrew, A.E. (2003) Public exposure: for better and for worse. *Museum International*, 50(4): 15–20.

Meskell, L. (2005) Sites of violence: terrorism, tourism and heritage in the archaeological present. In L. Meskell and P. Pels (eds), *Embedding Ethics: Shifting Boundaries of the Anthropological Profession*, pp. 123–46. London: Berg.

Millar, S. (1989) Heritage management for heritage tourism. *Tourism Management*, 10 (1): 9–14.

Mortensen, L. (2006) Structural complexity and social conflict: managing the past at Copan, Honduras. In D. Brodie (ed.), *Archaeology, Cultural Heritage and the Antiquities Trade*, pp. 258–69. Gainesville, FL: University Press of Florida.

Moscardo, G. (1996) Mindful visitors: heritage and tourism. *Annals of Tourism Research*, 23: 376–97.

Mowforth, M. and Munt, I. (1998) *Tourism and Sustainability: New Tourism in the Third World*. London: Routledge.

Ng, M.K. (2002) From a "cultural desert" to a "cultural supermarket": tourism promotion in Hong Kong. In W.B. Kim and J.Y. Yoo (eds), *Culture, Economy and Place: Asia-Pacific Perspectives*, pp. 179–218. Seoul: Korea Research Institute for Human Settlements.

Nyaupane, G.P., Morais, D.B., Dowler, L. (2006) The role of community involvement and number/type of visitors on tourism impacts: a controlled comparison of Annapurna, Nepal and Northwest Yunnan, China. *Tourism Management*, 27: 1373–85.

Pandey, R.N., Chettri, P., Kunwar, R.R. and Ghimire, G. (1995) *Case Study on the Effects of Tourism on Culture and the Environment: Nepal*. Bangkok: UNESCO.

Parnwell, M.J.G. (1998) Tourism and critical security, with particular reference to Burma. In N. Poku and D.T. Graham (eds), *Redefining Security: Population Movements and National Security*, pp. 123–48. Westport, CT: Greenwood.

Prott, L.V. (1996) Saving the heritage: UNESCO's action against illicit traffic in Africa. In P.R. Schmidt and R.J. McIntosh (eds), *Plundering Africa's Past*, pp. 29–44. Bloomington, IN: Indiana University Press.

Rogers, S.C. (2002) Which heritage? Nature, culture, and identity in French rural tourism. *French Historical Studies*, 25(3): 475–503.

Singh, T.V. and Singh, S. (2004) On bringing people and park together through ecotourism: the Nanda Devi National Park, India. *Asia Pacific Journal of Tourism Research*, 9(1): 43–55.

Smith, M. (2003) *Issues in Cultural Tourism Studies*. London: Routledge.

Smith, V. (ed.) (1989) *Hosts and Guests: the Anthropology of Tourism*. Philadelphia, PA: University of Pennsylvania Press.

Son, B.H. (2004) Tourism and the preservation of heritage sites in Vietnam: a case study of a water buffalo fighting festival and its tourist attraction. *Social Sciences*, 6: 31–44.

Steinberg, F. (1996) Conservation and rehabilitation of urban heritage in developing countries. *Habitat International*, 20(3): 463–75.

Tilley, C. (1997) Performing culture in the global village. *Critique of Anthropology*, 17 (1): 67–89.

Timothy, D.J. (1994) Environmental impacts of heritage tourism: physical and socio-cultural perspectives. *Manusia dan Lingkungan*, 11(4): 37–49.

—— (1999) Built heritage, tourism and conservation in developing countries: challenges and opportunities. *Journal of Tourism*, 4: 5–17.

—— (2000) Building community awareness of tourism in a developing country destination. *Tourism Recreation Research*, 25(2): 111–16.

Timothy, D.J. and Boyd, S.W. (2003) *Heritage Tourism*. Harlow: Prentice Hall.

Timothy, D.J. and Olsen, D.H. (eds) (2006) *Tourism, Religion and Spiritual Journeys.* London: Routledge.

Timothy, D.J. and Prideaux, B. (2008) Emerging issues and directions of cultural heritage tourism in the Asia Pacific region. In B. Prideaux, D.J. Timothy and K. Chon (eds), *Cultural and Heritage Tourism in Asia and the Pacific*, pp. 315–21. London: Routledge.

Timothy, D.J. and White, K. (1999) Community-based ecotourism development on the periphery of Belize. *Current Issues in Tourism*, 2(2/3): 226–42.

Urry, J. (2002) *The Tourist Gaze*, 2nd edn. London: Sage.

Wagner, J.E. (1996) Estimating the economic impacts of tourism. *Annals of Tourism Research*, 24: 592–608.

Wall, G. (1989) An international perspective on historic sites, recreation, and tourism. *Recreation Research Review*, 14(4): 10–14.

Wall, G. and Mathieson, A. (2006) *Tourism: Change, Impacts and Opportunities.* Harlow: Prentice Hall.

Wang, J. (2004) Claiming our heritage. *Beijing Review*, 47(25): 32–33.

Wang, Y. and Wall, G. (2005) Resorts and residents: stress and conservatism in a displaced community. *Tourism Analysis*, 10(1): 37–53.

Wilkie, D.S. and Carpenter, J. (2002) Can nature tourism help finance protected areas in the Congo Basin? *Oryx*, 33(4): 333–39.

Wood, R.E. (1993) Tourism, cultural and the sociology of development. In M. Hitchcock, V.T. King and M.J.G. Parnwell (eds), *Tourism in Southeast Asia*, pp. 48–70. London: Routledge.

Woods, L.A., Perry, J.M. and Steagall, J.W. (1994) Tourism as a development tool: the case of Belize. *Caribbean Geography*, 5(1): 1–19.

Part II
Heritage issues and challenges
Regional perspectives

5 The meanings, marketing, and management of heritage tourism in Southeast Asia

Joan C. Henderson

Introduction

Heritage is critical to tourism, motivating travelers and forming a basis for industry products and services, and as well as being a key component in destination marketing campaigns. It represents economic capital but also has a social and political value that affects the ways in which it is experienced, interpreted, and presented to audiences who comprise both tourists and residents. There is scope for conflict, and the relationship between heritage and tourism may be a troubled one, although tourism can encourage an appreciation of heritage among all relevant parties and promote its conservation.

The challenges of managing the heritage–tourism relationship and resolving any tensions are influenced, and perhaps compounded, in certain countries by more general circumstances; this chapter examines conditions in Southeast Asia. The region has a wealth of heritage resources that constitute tourist attractions while often serving additional purposes. However, the merits of approaches toward their use and management are debatable and illustrate the difficulties of securing a satisfactory balance among alternative perspectives and competing interests.

After an opening section that provides some background information about Southeast Asian countries and summarizes their overall tourism performance and the part played by heritage, separate dimensions of heritage are explored. The focus is on socio-cultural, colonial, wartime, and political heritage with discussion also of formally designated heritage sites and broader conservation issues. A final conclusion reviews the material and highlights key points. Heritage emerges as a core tourism asset that has excellent prospects in Southeast Asia, although formidable obstacles will have to be addressed and overcome if it is to be successfully conserved and sustainably managed with particular risks of neglect, over-exploitation, degradation, and politicization.

Tourism and heritage tourist attractions in Southeast Asia

Ten countries comprise the Association of South East Asian Nations (ASEAN) and exhibit diversity in terms of geography, history, and socio-economic profile as disclosed by the statistics in Table 5.1. Political systems range from the "small autocratic sultanate" of Brunei (EIU 2007a: 4), through professed parliamentary democracies of assorted credibility and competence, to Vietnam and Laos, which remain firmly communist despite market reforms. With the exception of Brunei and Singapore, most are relatively poor and have only attained "medium human development" status according to a UN index, but there are contrasts in circumstances suggested by the ranking of Malaysia and Laos at 61 and 133 respectively (UNDP 2005). The region is prone to instability, aggravated by endemic corruption and poverty in some nations. Such features, together with recent terrorist activity, natural disasters, and health scares, have impacted negatively on both tourism and whole economies (EIU 2006, 2007a–i).

Variations regarding tourism are revealed in Table 5.2, which lists international arrivals for 2006 by state, from under a million in Laos and Myanmar/ Burma to over 10 million in Malaysia and Thailand. While forecasted growth

Table 5.1 ASEAN member country statistics

Country	Land area (sq km)	Population (thousands)	Language	Religion	Per capita GDP (US$) 2006
Brunei	5,765	383	Malay	Islam	30,214
Cambodia	181,035	14,163	Khmer	Buddhism	512
Indonesia	1,919,317	222,051	Bahasa Indonesia	Islam Hinduism Buddhism Christianity	1,640
Lao PDR	236,800	6,135	Lao	Buddhism	574
Malaysia	330,113	26,686	Malay English Chinese Tamil	Islam Buddhism Christianity Hinduism Taoism	5,880
Myanmar	676,575	57,289	Burmese	Buddhism Christianity Islam	209
The Philippines	300,000	86,910	Filipino English Spanish	Christianity Islam Buddhism	1,351
Singapore	647.8	4,484	English Malay Mandarin Tamil	Buddhism Christianity Islam	29,500
Thailand	513,115	65,233	Thai	Buddhism Islam	3,168
Vietnam	331,700	84,222	Vietnamese	Buddhism Christianity	724

GDP, gross domestic product.
Source: ASEAN (2007).

rates are good (UNWTO 2007), disappointing volumes in certain instances hint at unrealized potential and the presence of substantial barriers to destination development. Most travel is intra-Asian, and details of the principal markets for ASEAN collectively are contained in Table 5.3, the pattern being repeated at a national level. Domestic tourism is expanding in scale as economies advance and travel opportunities are enlarged, but comprehensive figures are not available.

Joint marketing and product development is undertaken by an ASEAN Tourism Association (ASEANTA 2007) and a handful of regional tourism partnerships. One noteworthy example is the Agency for Coordinating Mekong Tourism Activities, a component of the wider Mekong economic program covering the territories through which the Mekong River flows and sponsored by the Asian Development Bank (ADB 2004). Member countries are Cambodia, Laos, Burma/Myanmar, Thailand, and Vietnam alongside the

Table 5.2 International tourist arrivals by ASEAN member country, 2006

Country	Arrivals (thousands)
Brunei*	127
Cambodia	1,700
Indonesia	2,694
Lao PDR	857
Malaysia	12,903
Myanmar	630
The Philippines	2,843
Singapore	9,673
Thailand	13,822
Vietnam	3,582
Total	48,831

*Figures for 2005.
Source: ASEAN (2007).

Table 5.3 ASEAN top ten tourist-generating markets, 2005

Country	Arrivals (thousands)	Share (%)
ASEAN	23,254	45
European Union	5,238	10
Japan	3,650	7
China	3,007	6
Republic of Korea	2,645	5
USA	2,306	4.5
Australia	2,034	4
Taiwan	1,605	3
India	1,240	2
Hong Kong SAR	1,022	2
Rest of the world	5,284	10
Total	51,288	

Source: ASEAN (2007).

Chinese province of Yunnan. Such institutions acknowledge the significance of heritage, which is prominent in their advertising, and have attempted to create touring circuits founded on the theme. Collaborative ventures are limited, and ASEAN members market themselves independently, but National Tourism Organizations again allocate a high priority to dimensions of heritage, as evidenced by their websites.

Brunei concentrates on Islamic and royal heritage and that of the Malays and indigenous peoples of Borneo (Brunei Tourism 2007), while Cambodia focuses on its Khmer origins symbolized by the temples of Angkor (Tourism Cambodia 2007). Cultural tourism, historical sites, and pilgrimage tours appear in Indonesian advertising (Ministry of Culture and Tourism 2007), and Laos also seeks to showcase its cultural heritage (Yamauchi and Lee 1999). Malaysia celebrates a heritage of mixed races (Tourism Malaysia 2007), and Myanmar/Burma boasts of a 2,000-year heritage (MTPB 2007). Philippine architectural heritage is highlighted (PTA 2007), and Singapore markets its ethnic quarters, museums, and Second World War sites (STB 2007). Culture is identified by the Thai authorities as a distinct product (TAT 2007), and "historical and cultural vestiges" are categorized as primary attractions in Vietnam (Vietnam Tourism 2007).

Heritage is thus central to the destination marketing of official agencies, as well as the excursions and vacation packages devised and distributed by tour operators and travel agents working at home and abroad. It is an important means of positioning, often employed in efforts to capture the more affluent overseas tourists who are believed to make up the culture and heritage tourism markets. Although attention has tended to be concentrated on longer haul Western visitors, Asian heritage tourists are now likely to outnumber non-Asians, and differences in expectations, demands, and impacts must be acknowledged (Winter 2007). Residents and domestic tourists are also attracted, and it is possible that enthusiasm about heritage and nostalgia for times past will intensify among the region's citizens as countries modernize, in conformity with Western trends. All tourists, however, occupy positions on a continuum of interest from casual to serious, and some classes of heritage are of greater significance than others.

Facets of heritage related to societies and cultures are perhaps at the forefront of tourist and industry interest, but historical events and eras and the ways in which these have shaped contemporary societies are not ignored. The special qualities of colonial, wartime, and political heritage in addition to that of a socio-cultural nature are analyzed in the following sections, which also explore the dilemmas inherent in their exploitation as attractions for tourists.

Socio-cultural heritage

The region has an array of cultures, races, and religions that hold a fascination for many visitors. The tourism industry has drawn on this richness to fashion attractions and experiences, of differing degrees of authenticity, for tourist consumption. However, certain characteristics are also a cause of discord within societies, which is at odds with the pictures of harmony and vibrancy painted and circulated by those involved in the business of tourism.

Many ASEAN countries are culturally homogeneous, such as Cambodia where 90 percent of citizens are Khmer, but the Chinese diaspora has resulted in sizeable Chinese communities across Southeast Asia. Colonialism also encouraged the inward migration of workers, demonstrated by the mixed racial composition of Malaysia and Singapore. Over 75 percent of Singaporeans are Chinese, the remainder being principally Malays and Indians (EIU 2007b) who are all represented in Malaysia where Malays make up 60 percent of the population (EIU 2007c). Both countries have attempted to foster tolerant multiculturalism and avoid the ethnic strife that is recurrent in Indonesia, where 4 million ethnic Chinese live in a society that is 95 percent Malay (EIU 2007d), although rivalries and resentments persist (Harper 1998; Lai 1995).

There are also numerous ethnic minorities in the region, especially in the more remote and upland areas or border zones, whose territories predate existing boundaries. Hill tribes in Thailand (EIU 2007e) and Laos (EIU 2006), and Vietnam's fifty-three ethnic groups (EIU 2007f), comprise sizeable proportions of the population. The Malaysian states of Sabah and Sarawak on Borneo are home to aboriginal tribes. Myanmar has at least 100 ethnic groups of different religious affiliations besides the dominant Burmese (EIU 2007g), and there are 300 minorities in Indonesia (EIU 2007c). Their villages, ways of life, and arts and crafts have become tourist products, but the overall treatment of minorities has prompted criticisms about discrimination and exclusion (Matthews 2001; New Frontiers 2005a). A few groups have been pressing for greater political recognition and autonomy, a move strongly rejected by governments that fear national fragmentation and loss of authority.

Religion may be closely allied to race or ethnicity and can be a strong national bond, as in Thailand where 95 percent of the inhabitants are Buddhists. However, religious divergences and misunderstandings may ignite domestic and international animosities. A divisive trend has been Islamic revivalism, accompanied by mounting extremism and the politicization of the religion. Malaysia and Indonesia are predominantly Muslim, but traditionally have been moderate and secular republics (Nagata 1994). Increased religious orthodoxy now informs political debate, particularly in Malaysia where the government is anxious to be seen as the defender of Islam in a bid to counter an opposing theocratic party. More colorful manifestations of religion such as some ceremonies and buildings are tourism resources, but developments in which dynamics of race, religion, and politics combine to provoke hostility between Muslims and non-Muslims have the capacity to damage tourism (Henderson 2003a).

It is unsurprising that the harsh facts of socio-cultural and religious antagonisms and undercurrents of unrest are overlooked or downplayed within a tourism context. Deviations in perspective indicate the distance that can separate the worlds of destination residents and their rulers and tourists and the tourism industry, an outcome of commercial dictates as well as the function of tourism as a channel for the dissemination of visions of how authorities would like their country to be regarded. Such propaganda may appear harmless but can mean the manipulation of socio-cultural heritage to further political objectives and the privileging of chosen groups while others are ignored or misrepresented.

Several Southeast Asian countries are still relatively young in their present configuration, albeit with roots in ancient civilizations, and confront imperatives of defining and establishing a national and cultural identity. Heritage narratives and relics can be channeled to this end by conveying a unifying sense of a shared past and future destiny. Feelings of belonging are crucial when many races coexist, helping to transcend ethnic allegiances and cultivate sentiments of unity and loyalty (Gradburn 1997; Hall 1995). Fractures in societies must therefore be concealed in selective readings of history and depictions of harmonious multiculturalism which rarely correspond entirely with the realities.

Politicization of socio-cultural heritage and related tourism occurs across the region (Hall and Oehlers 2000; Richter 1999) and is blatant in the case of Burma, the name of which was altered to Myanmar by the ruling junta. The initially isolationist military dictatorship has been ostracized by much of the international community because of human rights abuses, yet officials have opened the doors to visitors to combat unfavorable images of the country and government and to earn much needed foreign exchange (Henderson 2003b). Notions of culture and heritage perpetrated in official tourism literature are grounded in the purported principle of the pre-eminence of the Burmese majority and an appropriation of Buddhism, with the army professing guardianship of national heritage and championship of social cohesion (Philp and Mercer 1999).

The manner in which societies and cultures, including their religious practices, are presented to tourists and transformed into commodities by the process is a topic of wider applicability and controversy. Tourism is an agent of change, for both better and worse, and its consequences are likely to be more profound in relatively poor and remote communities (Cole 2007). Locals are not always hapless bystanders and may willingly participate in and gain from commercialization (Cohen 1989, 2002), but some activities, events, and sites are vulnerable to over-exploitation. People too are at risk, especially certain ethnic minorities, and the tourism industry has obligations concerning the dignified treatment of destination residents that it does not always seem ready to meet. At the same time, tourist predilection for distinctive cultural heritages can nourish them and bolster minority resistance to marginalization and eradication.

Colonial heritage

Southeast Asia has been subjected to the influence of colonial powers, only Thailand escaping occupation. The British Empire once incorporated Brunei, Malaysia, Myanmar/Burma, and Singapore. The Spanish and Americans had interests in the Philippines. Many of the islands of present-day Indonesia belonged to the Dutch East Indies, and Cambodia, Laos, and Vietnam were ruled as Indochina under the French. Japanese annexation of large tracts of the region during the Second World War can also be termed a period of colonization. Although imperial ties have been largely severed, it is worth noting that former colonizers are often key tourist markets of ex-colonies, and travel companies in these generators tap into romanticized ideas of bygone days in their marketing.

Colonialism has also left intangible and tangible legacies for erstwhile colonies with which their authorities have had to deal (King 1976; Yeoh 1996), and newly independent countries often opt to destroy or neglect any physical remains (Western 1985). The attractiveness for tourists of some colonial era urban landscapes and sites is now, however, a consideration in official decisions. The esthetics of colonial architecture and opportunities for adaptive reuse to create new leisure spaces and enterprises provide a strong rationale for retention, at least of the façades. These arguments are commonly heard in Singapore and, to a lesser extent, in Malaysia, where colonial structures and districts have been conserved in their original or modified condition and marketed as attractions (Henderson 2004). French-style architecture is a distinctive element in the cities of Cambodia, Laos, and Vietnam where villas and administrative offices in addition to renowned colonial hotels have metamorphosed into luxury hotel properties (Peleggi 2005).

Years of colonization have to be explained in public museums admitting tourists and locals. This is usually done within a framework of nationalism in accordance with prevailing dictates (Brown 1994). Conceptions of nationhood are articulated, supported by historical evidence, and happenings and artifacts stand for markers on the journey to independence (Pretes 2003). However, inaccuracies and myths may be engendered and perpetuated by temporary and permanent museum exhibitions and stories of the past told elsewhere. Heritage and heritage tourism can thus be a theater for "post-colonial dialogue over nationalism" (Cartier 1996: 51) and harnessed to support government agendas.

Wartime heritage

Contemporary armed aggression is likely to deter most tourists, but past wars can have an appeal to those with a broad or more specific and possibly personal interest (Gordon 1998; Smith 1996). The predicaments of managing wartime heritage as a visitor attraction (Tunbridge and Ashworth 1996; Uzzell 1989) are especially acute when occurrences are within living memory, exemplified by instances in Southeast Asia—the scene of fierce fighting during the Second World War between the Japanese and the British and their allies for control. Disengagement from empire in the post-war years was often violent, and the Indochinese territories were embroiled in a struggle for independence from the French in the late 1940s and 1950s and then against the United

States in the next two decades. The second conflict saw unprecedented bombing of Laos and Vietnam and precipitated internal strife, the repercussions of which continue to reverberate; Cambodia was also ravaged by the war and its aftermath.

Nevertheless, locations linked to war are frequently a component of tourist itineraries in which participants may be nationals of old aggressors. The Changi Museum in Singapore, a tribute to those who were interned during the Second World War in the Japanese camp of the same name, is popular among the Japanese, British, and Australians. The so-called "death railway," built by prisoners of war in northern Thailand at great cost to human life, also draws visitors from countries with or without an historical connection. Even sunken naval ships in the waters of the Philippines and Brunei and Eastern Malaysia yield tourism opportunities, although the primary purpose is diving and not heritage appreciation.

In the case of Vietnam, visitors are invited to battlefields and military installations, and commercial tours are marketed to former foreign combatants. The Vietcong supply trail along the Laos–Cambodia border dating from the American war is judged to be a novelty for tourists, as are the Cambodian minefields, although these are still dangerous and local guides are essential for travelers exploring off the beaten track (EIU 2007h). Concerns have, however, been voiced about the utilization of war and its interpretation for visitors in Vietnam (Henderson 2000; Mydans 1999) and Thailand (Peleggi 1996). These echo the wider debate about dark tourism or thanatourism (Lennon and Foley 2000; Seaton 1999) where the motivation is encountering death and the macabre at sites such as prisons (Strange and Kempa 2003) and death camps (Beech 2000) in addition to those connected to war (Wight and Lennon 2004). As sometimes happens at other types of heritage attractions, entertainment or political propaganda takes precedence over education and "mindfulness" (Moscardo 1996), over-commercialization and politicization leading to trivialization of the subject matter and the obscuring of historical truths.

Political heritage

Colonial rule and war can be defined as political heritage. Ideologies, regimes, and personalities too have their own heritage, which is accepted as raw material for tourist attractions. This is striking with regard to communist states, and Ho Chi Minh, one of the leaders of the revolutionary movement in Vietnam, is a revered figure there. Ho's mausoleum and modest home are stops on organized tours of the capital city and his birthplace is a museum. Communism, both then and now, is hailed as a political doctrine and practice worthy of celebration. Similar sentiments are espoused in Laos, which commemorated thirty years under communism by opening the Lao People's History Army Museum in 2005, dedicated to nationalism and the communist cause (*The Straits Times* 2005a).

Reminders of Cambodia's Khmer Rouge dictatorship led by Pol Pot, when an estimated 1.7 million or 20 percent of the population died between 1975 and 1979, are more contentious. The Prime Minister announced a scheme in 2001 to "turn all of the country's genocide sites into tourism offices," including Pol Pot's final home, with a search for corporate sponsors (BBC 2003). Although there seems to have been little progress, two such venues rival the country's primary attraction of Angkor Wat in terms of visitors. A Museum of Genocide in the capital exhibits instruments of torture, blood stains, and photographs of those detained and killed in the building where it is housed, while the Choeng Ek "killing fields" are mass graves of thousands executed, the skulls of 8,000 of these on public display in a glass case (Istvan 2003). Both strive to inform about the horrors of a brutal administration, rather than honor its achievements, although signs of commercial activity may leave the visitor somewhat uncomfortable.

Observers inside and outside of Cambodia have complained about how the country's political heritage is being treated as a commodity and warn of the probability of Khmer Rouge theme parks. Their unease is confirmed by news stories of the Khmer Rouge Experience Café which "serves up a slice of life under the Pol Pot regime" with a menu of "rice water and leaves" delivered by waitresses "dressed in the black fatigues worn by Pol Pot's guerrillas" (*The Straits Times* 2005b). Municipal plans to privatize the Choeng Ek site by granting a Japanese company a thirty-year lease to operate the memorial were, however, abandoned at the last minute. The company was to have paid an annual fee of US$15,000 and intended to build a visitor center, introduce charges for Cambodians, and raise the cost of admission for foreigners by 600 percent (BBC 2005).

Heritage designation

The most widely known designation universally is perhaps that endowed by UNESCO. Its World Heritage Sites (WHSs) in the region are recorded in Figure 5.1. The system is not without its critics who claim that the welfare of the least powerful stakeholders may be ignored in pursuit of the award and subsequently with few attempts to engage locals in procedures (Aas *et al.* 2005). The omission of residents from decisions taken about WHSs in their midst is commented on by Buergin (2003), who writes about a Thai natural WHS where there have been attempts to remove the villages of an ethnic minority that is judged to be a disruptive presence. Human displacement at cultural WHSs has also been observed in Thailand (Black and Wall 2001) and Indonesia (Hampton 2005).

Another drawback is the tendency of governments, destination marketers, and the tourism industry to see listing as a unique selling point and to energetically try to maximize ensuing commercial opportunities. Such activity surrounds Angkor Wat, which has become a tourism icon of Cambodia, and WHSs in Mekong countries are sold as "Jewels of the Mekong," with talks

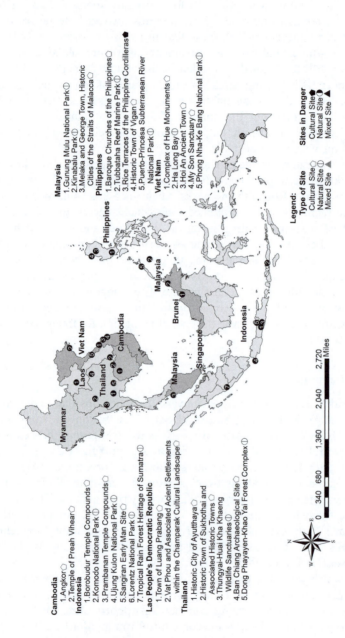

Cambodia
1. Angkor○
2. Temple of Preah Vihear○

Indonesia
1. Borobudur Temple Compounds○
2. Komodo National Park⊕
3. Prambanan Temple Compounds○
4. Ujung Kulon National Park⊕
5. Sangiran Early Man Site○
6. Lorentz National Park⊕
7. Tropical Rain Forest Heritage of Sumatra⊕

Lao People's Democratic Republic
1. Town of Luang Prabang○
2. Vat Phou and Associated Acient Settlements
   within the Champarak Cultural Landscape○

Thailand
1. Historic City of Ayutthaya○
2. Historic Town of Sukhothai and
   Associated Historic Towns○
3. Thungyai-Huai Kha Khaeng
   Wildlife Sanctuaries⊕
4. Ban Chiang Archaeological Site○
5. Dong Phayayen-Khao Yai Forest Complex⊕

Malaysia
1. Gunung Mulu National Park⊕
2. Kinabalu Park⊕
3. Melaka and George Town, Historic
   Cities of the Straits of Malacca○

Philippines
1. Baroque Churches of the Philippines○
2. Tubbataha Reef Marine Park⊕
3. Rice Terraces of the Philippine Cordilleras●
4. Historic Town of Vigan○
5. Puerto-Princesa Subterranean River
   National Park⊕

Viet Nam
1. Complex of Hue Monuments○
2. Ha Long Bay⊕
3. Hoi An Ancient Town○
4. My Son Sanctuary○
5. Phong Nha-Ke Bang National Park⊕

Legend:

Type of Site
Cultural Site ○
Natural Site ⊕
Mixed Site ▲

Sites in Danger
Cultural Site ●
Natural Site ●
Mixed Site ▲

0 340 680 1,360 2,040 2,720
Miles

Figure 5.1 Countries and World Heritage Sites in Southeast Asia

about a "Heritage Necklace" campaign that would incorporate all sites in single tour packages (New Frontiers 2005b). Intense coverage may boost visitation, yet this can have negative consequences (ICOMOS and WTO 1993; Shackley 1998). Popular sites suffer from the wear and tear of a constant stream of visitors (Ross 2005), and the more accessible ones are a magnet for beggars, hawkers, and guides whose unregulated trading can cause physical damage and mar the atmosphere.

Authorities have obligations to facilitate access to sites while demonstrating competence in protection, tasks that are not always easy to accomplish and reconcile (Drost 1996). There are doubts about the ability of some ASEAN governments, already handicapped by severe resource constraints, to exercise their responsibilities and effectively manage sensitive environments. Politicians may be apathetic or hostile, the overall planning system inadequate, and communication networks incomplete (Aas *et al.* 2005). Comprehensive guidelines and plans have been drawn up, often in cooperation with overseas consultants (Wager 1995), but these may be too general and overly ambitious, which renders full implementation unrealistic.

The vulnerability of UNESCO sites as a whole is reflected in the World Heritage in Danger label, applied when the special qualities that led to the original inscription are at risk from "armed conflict and war, earthquakes and other natural disasters, pollution, poaching, uncontrolled urbanisation and unchecked tourist development." Further hazards are the extreme heat and humidity of the tropical climate, dense and destructive vegetation, and looting. Labeling allows funding to be released and can be a catalyst for improved conservation efforts. The Philippine rice terraces were added to the list in 2001, but Angkor Wat was removed in 2004 after moves to abate problems of unauthorized excavation, pillaging, and landmines (World Heritage Centre 2007b). The World Monuments Fund (WMF) also has a "most endangered" register of 100 monuments, intended to stimulate awareness and prompt rescue, one of which is the Kotagede Historic District on the Indonesian island of Java (World Monuments Fund 2007).

Not all valuable heritage enjoys the safety afforded by formal international recognition, and lack of protective designation at national and local level exposes assets to damage and destruction. Kampong Cina river frontage in Kuala Terengganu, Malaysia, was targeted for preservation by the WMF because of proposals to raze the whole village. The area was deemed unsightly and an impediment to tourism by officials even though it accommodated every phase in the evolution of the shophouse—an architectural design that marries working and living quarters and is typical of the region (*The Straits Times* 1998). Unsympathetic government attitudes and action, or inaction, underlie another controversial redevelopment scheme that has already fundamentally altered the historic riverside of Malacca, Malaysia (*The New Straits Times* 2005). In addition, countless other sites throughout the region are in jeopardy.

Issues of heritage conservation

As suggested in the above sections, Southeast Asia has a diverse and highly distinctive heritage that merits recognition and saving for the benefit of both tourists and residents. However, there are several barriers to successful conservation. Many of these are common to the much of the developing world, but the particular conditions prevailing regionally and nationally add a unique dimension to the issues surrounding the conserving of heritage and sustainable tourism as a whole.

Sustainable tourism implies a commitment to heritage conservation on the part of the industry, which can contribute through financial support and the stewardship of heritage resources within its ambit or where there is shared control, but this may be lacking. Funding and strategic planning are also essentially public sector duties. Reference has already been made to some of the difficulties confronting Southeast Asian governments. Insufficient money and technical skills and competing land uses handicap conservation work together with unfettered tourism. Even if attempts are made, the results can be disappointing as in Bagan in Myanmar, where new pagodas have been built on top of ancient ruins using inauthentic materials in a "false and misguided restoration" (CNN 2003; New Frontiers 2005c).

Nevertheless, sufficient investment and proper techniques can help safeguard physical heritage; in comparison, more intangible aspects are harder to retain. Ways of life may be eroded by modernization, and a society's customs, crafts, festivals, and other traditions become diluted or disappear over time. Harmful socio-cultural change can be introduced and accelerated by undesirable and preventable practices such as land appropriation and illegal logging, which disrupt the lives of residents. Official ethnic minority programs can frequently lead to assimilation or displacement, preventing peoples from pursuing long-established modes of living. Private and public tourism projects may also require the disturbance or even shifting of communities. Tourism itself is both an outcome and an instrument of globalization.

The Karen women of Thailand, refugees from neighboring Myanmar and famed for the brass coils worn around their elongated necks, are often cited as a striking example of the exploitation and abuse of disadvantaged minorities. Thai entrepreneurs have sought to capitalize on their novelty, and tour operators are alleged to have enticed Karen groups fleeing cross-border violence and detained them in camps before they are put on display. Despite criticism of the treatment of the "long-necked" women as captives in a human zoo, visits to their villages are highlighted in the tourism marketing of Thailand's northern provinces and remain popular with tourists (Moe 2005).

Elsewhere, rapid development and the ensuing socio-cultural losses are bemoaned by some commentators, while others welcome them as signs of progress, enhancing the quality of life for residents. Heritage protection extending beyond edifices and artifacts may be justified, but societies are robust and "each generation redefines its heritage in response to new understandings, new experiences and new inputs from an ever-increasing range of contacts from outside" (Sofield 2000: 51). UNESCO's classification of the Philippino Ifugao ethnic chant as one of the "Masterpieces of the Oral and Intangible Heritage of Humanity" (World Heritage Centre 2007a) and the preparation of a preservation master plan is censured for failing to appreciate these dynamics. Reyes (2006) contends that the indigenous culture embodied in musical forms is not static and local people should be free to express themselves without deferring to an outside elite about what is worthy.

Tourists too can have romanticized and idealized expectations of societies and cultural experiences, engaging in quests for the "primitive," which are fostered by the tourism industry (Adams 1984). Nostalgia is also felt by residents, and the rapid transformation of Asian societies and landscapes has prompted reflections on memory and identity (Chang 2005), especially among the emerging middle classes, which has consequences for popular interest in and the official priority attached to identifying and saving heritage (Bradford and Lee 2004).

There may be an unwillingness to tolerate the intrusion of modernity in some environments, giving rise to reconstructions in which host populations collude. Longhouse inhabitants of the Iban tribe in Borneo, for example, maintain their traditional dwellings as tourist attractions and accommodation while living permanently in more comfortable and updated variations that are rejected by tour operators and visitors as insufficiently "ethnic" (Zeppel 1997). Even in the developed destination of Singapore, socio-cultural representations are colored and modified to satisfy commercial demands (Leong 1997). A degree of play acting and creativity may be acceptable, but heritage is debased when conservation is overridden by manufacture and invention, and heritage tourism is reduced to a trade in falsehoods and contrivances.

Questions of authenticity have inspired much discussion in the tourism literature in the past (Cohen 1979; MacCannell 1976; Pearce and Moscardo 1986) and continue to do so (Chhabra *et al.* 2003; Steiner and Reisinger 2006; Yeoman *et al.* 2007). The concept is elusive and disputed, different interpretations complicating satisfactory definitions. Wang (1999) maintains that there are three types of authenticity – original, which describes original items; constructive, which is transmitted to things and experiences by tourists and the tourism industry; and existential, which refers to that resulting from tourist involvement. Tourists have been shown to be satisfied with what experts might dismiss as fake, and the industry to deal in images and idealized depictions of people and places in a post-modern world where realities have become unclear and ambiguous.

These general issues have a resonance in Southeast Asia (Teo *et al.* 2001), but heritage conservation is further affected by the political and socio-economic instability present in much of the region. Locations believed to be unstable for whatever reason will be avoided by most tourists, the tourism industry in generating markets, and investors who are all very risk averse. Tourism overall is impeded, including attendance at heritage sites with lost revenues and possibly less funding for conservation. Attitudes toward heritage could also be influenced by instability as authorities and populations are preoccupied with immediate crises engendered by turbulence.

Likewise, corruption appears endemic in many administrations, and there is evidence of cronyism among leaders who fail to discharge their public responsibilities. Regimes of dubious legitimacy hamper effective government, economic progress, and planning, resulting in negative repercussions for policy-making and the safeguarding of heritage. Security risks are heightened and law and order may be weak, exemplified by the widespread looting at archeological sites in Thailand and Cambodia and the illicit trade in antiquities (Thosarat 2001). The pervasive poverty in the least developed countries of the region exacerbates the problems of good government, incorporating heritage strategies, and formidable barriers to general development are in operation in the poorest states of Cambodia and Laos. Inadequate communications infrastructure and investment and an overdependence on foreign aid are other features that impact on heritage conservation as well as the devising and execution of practical programs.

Even if there are formal plans founded on a sustainable approach intended to avoid depletion of, and damage to, heritage resources, legislation may not be fully enforced and administrative mechanisms may be absent. Tools such as participatory planning are sometimes alien concepts, and enthusiasm for short-term gains among governments may contradict sustainability objectives. Heritage can be a valuable source of private and public revenue, and proper protection is made harder in desperately poor countries where tourism may be one of the few ways of earning a living or supplementing low incomes for beggars, orphans, souvenir hawkers, refreshments vendors, and guides who congregate at sites frequented by tourists.

Alongside political and socio-economic considerations, dangers to heritage originate in the forces of nature. Certain areas are prone to natural disasters. For example, the tsunami at the end of 2004 caused widespread devastation around the shores of the Indian Ocean and an estimated 223,000 deaths including many in Indonesia and Thailand. Although the example is unprecedented in its scale, tropical storms and flooding are recurrent in Southeast Asia, parts of which are also vulnerable to earthquakes and volcanic eruptions. These events have the ability to inflict severe damage, especially when defenses are inadequate and governments inept or ill-equipped to respond, and threaten living and other heritage. The constant threats and actual eruptions of volcanoes on the islands of Bali and Java threaten several important heritage sites, including some on UNESCO's World Heritage List such as, Borobudur and Prambanan.

Conclusion

Each country in Southeast Asia is unique and possesses its own history and heritage, which have the power to entice visitors to single or multiple destinations. The allure of heritage is revealed in the range of attractions based upon it, their inclusion in individual itineraries and organized tours, and the space allocated to the topic in destination marketing. Heritage takes many forms, a selection of which have been discussed in this chapter, and tourism has made use of unlikely manifestations related to political regimes and war, as well as devising more conventional products from social and cultural expressions.

In addition to its commercial functions, heritage has been shown to have other important purposes that affect how it is presented to and received by audiences at home and abroad. Heritage is a type of social capital and a repository of memories to be handed on in good condition to subsequent generations if societies are to understand their origins and history. Governments have a vital part to play as custodians of heritage and facilitators of explications of the meanings of history, but the role presumes an objectivity that may be compromised by political ideologies and agendas. Heritage can therefore additionally act as political capital to be expended in pursuit of broader policies, often aligned to hegemonic and economic goals, and especially nation-building.

Heritage encompasses views of the past and its contemporary relevance is filtered through the lens of the present and possibly distorted by economic, political, and socio-cultural pressures. Meanings and applications are fluid and depend upon individuals, groups, governments, and international organizations with scope for disagreement within and among constituencies. Such interactions and the tensions engendered are evident in the arena of tourism, and the relationship between heritage and tourism can be difficult in Southeast Asia, as in the rest of the world, where contentious issues of interpretation and presentation must be tackled alongside practical matters.

Meeting the challenges of effective conservation may mean domestic reforms to structures and processes and seeking assistance from abroad with regard to finance, expertise, and equipment. Heritage managers must also contend with official economic development policies in which the highest priority is given to growth and revenue generation with less enthusiasm for recalling and preserving the past. The right to make money from heritage cannot be denied, and conservation does not preclude commercialization, provided an appropriate infrastructure is installed and a sustainable approach is adopted. However, reaching acceptable compromises is made harder in desperately poor countries such as some ASEAN members.

Future success in heritage management depends upon addressing these questions and advancing toward the resolution of urgent problems. Progress is also critical to tourism as the industry relies upon the national and local distinctiveness captured and conveyed in heritage and a stock of well-preserved resources and high-quality attractions. Destinations without such assets are at a competitive disadvantage, and there would appear to be a coincidence of commercial and non-commercial interest concerning the survival of the region's heritage. Tourism can thus be a benefactor of heritage, as well as a threat that requires containment, and its positive contribution should not be forgotten.

References

Aas, C., Ladkin, A. and Fletcher, J. (2005) Stakeholder collaboration and heritage management. *Annals of Tourism Research*, 32(1): 28–48.

Adams, K. (1984) Come to Tana Toraja, "Land of the Heavenly Kings": travel agents as brokers in authenticity. *Annals of Tourism Research*, 11(3): 469–85.

ADB (2004) *Technical Assistance for the Greater Mekong Subregion Tourism Sector Strategy*. Manila: Asia Development Bank.

ASEAN (2007) ASEAN Statistics. ASEAN. Available from http://www.aseansec.org (accessed September 13, 2007).

ASEANTA (2007) About ASEANTA. ASEANTA. Available from http://www.aseanta.org (accessed September 13, 2007).

Beech, J. (2000) The enigma of holocaust sites as tourist attractions: the case of Buchenwald. *Managing Leisure*, 5: 29–41.

Black, H. and Wall, G. (2001) Global–local interrelationships in UNESCO World Heritage Sites. In P. Teo, T.C. Chang and H.K. Chong (eds), *Interconnected Worlds: Tourism in Southeast Asia*, pp. 121–36. Amsterdam: Elsevier.

Bradford, M. and Lee, E. (eds) (2004) *Tourism and Cultural Heritage in Southeast Asia*. Bangkok: SEAMEO Regional Centre for Archaeology and Fine Arts.

BBC (2003) Cambodia cashes in on grim past. BBC News, September 12. Available from http://www.newsvote.bbc.co.uk (accessed January 28, 2006).

—— (2005) Killing Fields deal hits delay. BBC News, April 7. Available from http://www.newsvote.bbc.co.uk (accessed January 28, 2006).

Brown, D. (1994) *The State and Ethnic Politics in South East Asia*. New York: Routledge.

Brunei Tourism (2007) Heritage. Available from http://www.tourismbrunei.com (accessed September 14, 2007).

Buergin, R. (2003) Shifting frames for local people and forests in a global heritage: the ThungYai Naresuan Wildlife Sanctuary in the context of Thailand's globalisation and modernisation. *Geoforum*, 34: 375–93.

Cartier, C. (1996) Conserving the built environment and generating heritage tourism in Peninsular Malaysia. *Tourism Recreation Research*, 21(1): 45–53.

Chang, T.C. (2005) Place, memory and identity: imaging "New Asia". *Asia Pacific Viewpoint*, 46(3), 247–53.

Chhabra, D., Healy, R. and Sills, E. (2003) Staged authenticity and heritage tourism. *Annals of Tourism Research*, 30(3): 702–19.

CNN (2003) Outcry over Burma temple tower. CNN. November 30. Available from http://www.cnn.com (accessed January 28, 2006).

Cohen, E. (1979) Rethinking the sociology of tourism. *Annals of Tourism Research*, 6: 18–35.

—— (1989) Primitive and remote: hill tribe trekking in Thailand. *Annals of Tourism Research*, 16: 30–61.

—— (2002) *The Commercialised Crafts of Thailand: Hill Tribes and Lowland Villages*. Andover: Curzon.

Cole, S. (2007) *Tourism, Culture and Development: Hopes, Dreams and Realities in East Asia*. Clevedon: Channel View Publications.

Drost, A. (1996) Developing sustainable tourism for World Heritage Sites. *Annals of Tourism Research*, 23(2): 479–92.

EIU (2006) *Country Profile 2005: Laos*. London: The Economist Intelligence Unit.

—— (2007a) *Country Profile 2007: Brunei*. London: The Economist Intelligence Unit.

—— (2007b) *Country Profile 2007: Singapore*. London: The Economist Intelligence Unit.

—— (2007c) *Country Profile 2006: Malaysia*. London: The Economist Intelligence Unit.

—— (2007d) *Country Profile 2007: Indonesia*. London: The Economist Intelligence Unit.

—— (2007e) *Country Profile 2007: Thailand*. London: The Economist Intelligence Unit.

—— (2007f) *Country Profile 2006: Vietnam*. London: The Economist Intelligence Unit.

—— (2007g) *Country Profile 2006: Myanmar*. London: The Economist Intelligence Unit.

—— (2007h) *Country Profile 2007: Cambodia*. London: The Economist Intelligence Unit.

——(2007i) *Country Profile 2006: The Philippines*. London: The Economist Intelligence Unit.

Gordon, B.M. (1998). Warfare and tourism. *Annals of Tourism Research*, 25(3): 616–38.

Gradburn, N.H. (1997) Tourism and cultural development in East Asia and Oceania. In Y. Yamashita, K.H. Din and J.S. Eades (eds), *Tourism and Cultural Development in Asia and Oceania*, pp. 194–214. Kuala Lumpur: University Kebangsaan Malaysia.

Hall, C.M. and Oehlers, A.L. (2000) Tourism and politics in South and Southeast Asia: political instability and policy. In C.M. Hall and S. Page (eds), *Tourism in South and Southeast Asia: Issues and Cases,* pp. 77–93. Oxford: Butterworth Heinemann.

Hall, S. (1995) New cultures for old. In D. Massey and P. Jess (eds), *A Place in the World? Places, Cultures and Globalisation*, pp. 175–213. Milton Keynes: Open University Press.

Hampton, M. (2005) Heritage, local communities and economic development. *Annals of Tourism Research*, 32(3): 735–59.

Harper, T.N. (1998) *The End of Empire and the Making of Malaysia*. New York: Cambridge University Press.

Henderson, J.C. (2000) War as a tourist attraction: the case of Vietnam. *International Journal of Tourism Research*, 2: 269–80.

—— (2003a) Managing tourism and Islam in Peninsular Malaysia. *Tourism Management*, 24: 447–56.

—— (2003b) The politics of tourism in Myanmar. *Current Issues in Tourism*, 6(2): 97–118.

—— (2004) Tourism and British colonial heritage in Malaysia and Singapore. In C.M. Hall and H. Tucker (eds), *Tourism and Postcolonialism: Contested Discourses, Identities and Representations*, pp. 113–25. London and New York: Routledge.

ICOMOS and WTO (1993) *Tourism at World Heritage Cultural Sites: The Site Manager's Handbook*. Madrid: International Council on Monuments and Sites and World Tourism Organisation.

Istvan, Z. (2003) Killing fields lure tourists in Cambodia. National Geographic News. Available from http://news.nationalgeographic.com (accessed August 25, 2004).

King, A.D. (1976) *Colonial Urban Development: Culture, Social Power and Environment*. London: Routledge and Kegan Paul.

Lai, A.E. (1995) *Meanings of Multiethnicity: A Case Study of Ethnicity and Ethnic Relations in Singapore*. Oxford and Singapore: Oxford University Press.

Lennon, J.J. and Foley, M. (eds) (2000) *Dark Tourism: The Attraction of Death and Disaster*. London: Continuum.

Leong, W.T. (1997) Commodifying ethnicity: state and ethnic tourism in Singapore. In M. Picard and R. Wood (eds), *Tourism, Ethnicity and the State in Asian and Pacific Societies*, pp. 71–98. Honolulu, HI: University of Hawaii Press.

MacCannell, D. (1976) *The Tourist*. New York: Schocken Books.

Matthews, B. (2001) Ethnic and religious diversity: Myanmar's ripening nemesis. Paper given at Institute of South East Asian Studies, National University of Singapore, March 29.

Ministry of Culture and Tourism (2007) Indonesia: the ultimate in diversity. Ministry of Culture and Tourism. Available from http://my-indonesia.info (accessed September 14, 2007).

Moe, K.Z. (2005) Tradition or sideshow? *New Frontiers*, 11(4): July–August.

Moscardo, G. (1996) Mindful visitors: heritage and tourism. *Annals of Tourism Research*, 3(2): 376–97.

MTPB (2007) Welcome to mystical Myanmar. Myanmar Tourism Promotion Board. Available from http://www.myanmar-tourism.com (accessed September 14, 2007).

Mydans, S. (1999) Visit the Vietcong's world: Americans welcome. *The New York Times*, July 7.

Nagata, J. (1994) How to be Islamic without being an Islamic state: contested models of development. In A.S. Ahmed and H. Donnan (eds), *Indian Communities in Southeast Asia*, pp. 513–40. Singapore: ISEAS and Times Academic Press.

New Frontiers (2005a) Government treatment of hill peoples deplored. *New Frontiers*, 11(4), July–August.

—— (2005b) Heritage necklace plan attracts criticism. *New Frontiers*, 11(5), September–October.

—— (2005c) Misguided restoration destroys Pagan. *New Frontiers*, 11(5), September–October.

The New Straits Times (2005) Sprucing up Malacca River. *The New Straits Times*, January 26.

Pearce, P.L. and Moscardo, G.M. (1986) The concept of authenticity in tourist experiences. *The Australian and New Zealand Journal of Sociology*, 22, 121–32.

Peleggi, M. (1996) National heritage and global tourism in Thailand. *Annals of Tourism Research*, 23(12): 432–48.

—— (2005) Consuming colonial nostalgia: the monumentalism of historic hotels in urban South-East Asia. *Asia Pacific Viewpoint*, 46(3), 255–65.

PTA (2007) Beautiful and exciting sites and trips. Philippine Tourism Authority. Available from http://www.philtourism.gov.ph (accessed September 14, 2007).

Philp, J. and Mercer, D. (1999) Commodification of Buddhism in contemporary Burma. *Annals of Tourism Research*, 26(1): 21–51.

Pretes, M. (2003) Tourism and nationalism. *Annals of Tourism Research*, 30(1): 125–42.

Reyes, M.Y. (2006) Philippines: irony of "dying tradition" narratives: Ifugao chant as UNESCO "masterpiece". Third World Network Clearinghouse. Available from http://www.twnside.org (accessed January 28, 2006).

Richter, L. (1999) The politics of heritage tourism: emerging issues for the new millennium. In D.G. Pearce and R.W. Butler (eds), *Tourism Development: Contemporary Issues*, pp. 108–26. London: Routledge.

Ross, D. (2005) Who benefits from Angkor tourism? *New Frontiers*, 11(2) March–April.

Seaton, A.V. (1999) War and thanatourism: Waterloo 1815–1914. *Annals of Tourism Research*, 26(1): 130–58.

Shackley, M. (ed.) (1998) *Visitor Management: Case Studies from World Heritage Sites*. Oxford: Butterworth Heinemann.

STB (2007) What to see. Uniquely Singapore. Available from http://www.visitsingapore.com (accessed September 14, 2007).

Smith, V. (1996) War and its tourist attractions. In A. Pizam and Y. Mansfeld (eds), *Tourism, Crime and International Security Issues*, pp. 247–64. Chichester: John Wiley and Sons.

Sofield, T. (2000) Rethinking and reconceptualising social and cultural issues in Southeast and South Asian tourism development. In C.M. Hall and S. Page (eds),

Tourism in South and Southeast Asia: Issues and Cases, pp. 45–57. Oxford: Butterworth Heinemann.

Steiner, C.J. and Reisinger, Y. (2006) Understanding existential authenticity. *Annals of Tourism Research*, 33(2): 299–318.

The Straits Times (1998) Asia's crumbling monuments. *The Straits Times*, February 24, p. 16.

—— (2005a) Communist Laos turns 30 today. *The Straits Times*, December 2.

—— (2005b) Pol Pot menu. *The Straits Times*, October 1.

Strange, C. and Kempa, M. (2003) Shades of dark tourism: Alcatraz and Robben Island. *Annals of Tourism Research*, 30(2), 386–405.

Teo, P., Chang, T.C. and Ho, K.C. (eds) (2001) *Interconnected Worlds: Tourism in Southeast Asia*. Oxford: Elsevier.

Thosarat, R. (2001) Report from Southeast Asia. *Culture Without Context*, 8 (Spring). Available from http://www.mcdonald.cam.ac.uk/projects/iarc/culturewithoutcontext/issue8/thosarat.htm (accessed December 17, 2007).

TAT (2007) Majestic cultural heritage. Tourism Authority of Thailand. Available from http://www.tourismthailand.org (accessed September 14, 2007).

Tourism Cambodia (2007) Culture. Tourism Cambodia. Available from http://www.tourismcambodia.com (accessed September 14, 2007).

Tourism Malaysia (2007) Culture and heritage. Tourism Malaysia, available from http://www.tourism.gov.my (accessed September 14, 2007).

Tunbridge, J.E. and Ashworth, G.J. (1996) *Dissonant Heritage*. Chichester: John Wiley and Sons.

UNDP (2005) *Human Development Report*. New York: United Nations Development Programme.

UNWTO (2007) *World Tourism Barometer*. Madrid: United Nations World Tourism Organisation.

Uzzell, D.L. (1989) *Heritage Interpretation*. Vol. 1. London and New York: Belhaven Press.

Vietnam Tourism (2007) Tourism. Vietnam Tourism. Available from http://www.vietnamtourism.com (accessed September 14, 2007).

Wager, J. (1995) Developing a strategy for the Angkor World Heritage Site. *Tourism Management*, 16(7): 515–23.

Wang, N. (1999) Rethinking authenticity in tourism experience. *Annals of Tourism Research*, 26(2): 349–70.

Western, J. (1985) Undoing the colonial city? *The Geographical Review*, 75(3): 335–57.

Wight, A.C. and Lennon, J.J. (2004) Towards an understanding of visitor perceptions of "dark" sites: the case of the Imperial War Museum in Manchester. *Journal of Hospitality and Tourism*, 2(2): 105–22.

Winter, T. (2007) Rethinking tourism in Asia. *Annals of Tourism Research*, 34(1): 27–44.

World Heritage Centre (2007a) World Heritage Site List. Available from http://whc.unesco.org (accessed September 14, 2007).

—— (2007b) World Heritage in Danger. Available from http://whc.unesco.org (accessed September 14, 2007).

World Monuments Fund (2007) World Monuments Watch. Available from http://www.wmf.org (accessed September 14, 2007).

Yamauchi, S. and Lee, D. (1999) *Tourism Development in the Lao People's Democratic Republic*. DESA Discussion Paper 9. New York: United Nations Department of Economic and Social Affairs.

Yeoh, B. (1996) *Contesting Space: Power Relations and the Urban Built Environment in Colonial Singapore.* Kuala Lumpur: Oxford University Press.

Yeoman, I., Brass, D. and McMahon-Beattie, U. (2007) Current issues in tourism: the authentic tourist. *Tourism Management*, 28(4): 1128–38.

Zeppel, H. (1997) Meeting "wild people": Iban culture and longhouse tourism in Sarawak. In S. Yamashita, K.H. Din and J.S. Eades (eds), *Tourism and Cultural Development in Asia and Oceania*, pp. 119–40. Bangi: Penerbit Universiti Kebang-saan Malaysia.

6 Heritage and tourism in East Asia's developing nations

Communist–socialist legacies and diverse cultural landscapes

Dallen J. Timothy, Bihu Wu, and Oyunchimeg Luvsandavaajav

Introduction

For tourism, East Asia is a unique region of the world for several reasons. First, it forms a mix of developed and developing countries, with the more affluent countries of the region being only relatively recently developed. Second, it is culturally diverse with a large variety of ethnic groups in only five states: China (People's Republic of China, including Hong Kong and Macao semi-autonomous regions (SARs) and Taiwan), Mongolia, Japan, North Korea, and South Korea. Third, while the countries in this region are few in number, they are among the most populated on earth. The region is geographically very large and home to nearly a quarter of the earth's population (approximately 23.5 percent). Fourth, the three less-developed countries of the region (North Korea, China, and Mongolia) have all been communist states and are all in various transitional stages of political change. Finally, the appeal of tourism in all three developing states lies in their heritage resources; in essence, tourism in East Asia is heritage tourism (including a very important natural heritage in China). Indicative of this is the high number of UNESCO-designated World Heritage Sites shown in Figure 6.1.

As noted earlier in Chapter 3, heritage is highly political, and many parties vie for power in any given location. Also mentioned in Chapter 3, Kim *et al.* (2007) and Timothy and Boyd (2003) identify several ways in which tourism is manipulated by the powerful to achieve some political end. The most pertinent for East Asia are: 1) when a country uses tourism as a tool for spreading propaganda to foreign visitors and extolling the virtues of a certain national ideology; and 2) using heritage and tourism to build nationalism and patriotism within a country's own citizenry. These two political uses are especially endemic to socialist–communist states, and the three developing countries (including China) in this region are no exception. Each of the three countries demonstrates this at varying degrees or levels. North Korea, for example, is the strictest communist regime remaining in the world today and has been since the 1950s. Tourism is strictly controlled and relatively few people travel in or out. When

China

1. Ancient Building Complex in the Wudang Mountains ◇
2. Ancient City of Ping Yao ◇
3. Ancient Villages in Southern Anhui ◇
4. Capital Cities and Tombs of the Ancient
   Koguryo Kingdom ◇
5. Classic Gardens of Suzhou ◇
6. Dazu Rock Carvings ◇
7. Fujian Tulou ◇
8. Historic Center of Macao ◇
9. Historic Ensemble of the Potala Palace ◇
10. Huanglong Scenic and Historic Interest Area ⊕
11. Imperial Palaces of the Ming and Qing Dynasties
    in Beijing and Shenyang ◇
12. Imperial Tombs of the Ming and Qing Dynasties ◇
13. Jiuzhaigou Valley Scenic and Historic Interest area ⊕
14. Kaiping Diaolou and Villages ◇
15. Longmen Grottoes ◇
16. Lushan National Park ◇
17. Mausoleum of the First Qin Emperor ◇
18. Mogao Caves ◇
19. Mount Emei Scenic Area, including Leshan Giant
    Buddha Scenic Area ▲
20. Mount Qingcheng and the Dujiangyan Irrigation System ◇
21. Mount Sanqingshan National Park ⊕
22. Mount Taishan ▲
23. Mount Wuyi ▲
24. Mountain Resort and its Outlying Temples, Chengde ◇
25. Old Town of Lijiang ◇
26. Sichuan Giant Panda Sanctuaries - Wolong,
    Mt Siguniang and Jiajin Mountains ⊕
27. South China Karst ⊕
28. The Great Wall ◇
29. Three Parallel Rivers of Yunnan Protected Areas ⊕
30. Wulingyuan Scenic and Historic Interest Area ⊕
31. Yin Xu ◇
32. Yungang Grottoes ◇
33. Mount Huangshan ▲
34. Peking Man Site at Zhoukoudian ◇
35. Summer Palace, an Imperial Garden in Beijing ◇
36. Temple of Heaven: an Imperial Sacrificial Altar in Beijing ◇
37. Temple and Cemetery of Confucius and the
    Kong Family Mansion in Qufu ◇

Mongolia

1. Orkhon Valley Cultural Landscape ◇
2. Uvs Nuur Basin ⊕

North Korea

1. Complex of Koguryo Tombs ◇

Legend:

Type of Site
Cultural Site ◇
Natural Site ⊕
Mixed Site ▲

Sites in Danger
Cultural Site ●
Natural Site ⊖
Mixed Site ▲

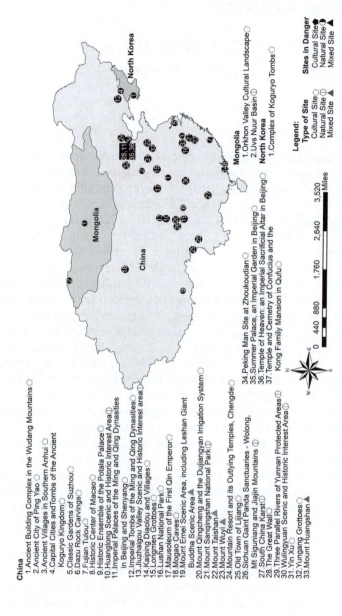

Figure 6.1 Developing Countries and World Heritage Sites of East Asia

foreign tourists do visit, they are required to be with a guide at all times, and the main destinations are related to nationalist ideals and glorifying the revolutionary past. This will be highlighted later in the discussion of North Korea.

China, while officially a communist state, functions economically as a capitalist society. Before the widespread growth of tourism in China during the 1990s, the country resembled closely the situation noted above; tourism was seen as a propaganda tool to illustrate the superiority of the communist system (Guangrui 1989; Hall 2001). Today, however, visitors are not required to visit nationalist sites or participate in tours of schools and factories that reinforce socialist ideals. Tourism in Mongolia prior to the 1990s followed the same pattern noted above, with the focus of the strictly controlled sector being schools, urban centers developed by the Soviets, and patriotic monuments to Soviet and Mongolian national heroes. Today, however, as a developing parliamentary republic, the country has shunned much of its communist past and moved toward a more globalized economy, including tourism. Like China, there is no longer a system in place to attempt to indoctrinate visitors about the virtues of a specific political system. This state-socialist past plays a very important role in all three countries' heritage tourism sectors, particularly still in North Korea.

Living culture and ethnic heritage are important ingredients in the tourism mix of the region as well. This is particularly the case in Mongolia, which is a fairly homogeneous country but with an interesting traditional culture, and China, a large country with dozens of ethnic minority groups, each trying to become involved in showing their heritage to Chinese and foreign visitors. Similarly, the region's ancient cultures and their material remains form a significant foundation for a thriving heritage tourism sector. This chapter aims to describe several of the main themes and issues in heritage tourism in the developing countries of East Asia (i.e., China, Mongolia, and North Korea), including the ways in which the political past, as noted above, and the cultural diversity influence the heritage product. The chapter also explains some of the management constraints and challenges being faced by the region.

Trends in heritage tourism in China

The People's Republic of China is demographically the largest country in the world and the fourth largest physically. The country's immensity traverses a vast array of physiographic regions, natural landscapes, and climatic zones, and is home to numerous ethnic and cultural groups. This diverse physical and cultural milieu underlies one of the richest resource bases for tourism on earth. China also has one of the oldest recorded human heritages, which it draws upon for much of its cultural appeal (Chang 1991; Gernet 1996).

China is often referred to as a communist state with a capitalist economy. Similar systems currently exist in Vietnam and Cambodia. China is a rapidly developing country and, since its recent admission to the World Trade Organization, international ties and trade have been strengthened further and growth is expected to continue accelerating. In terms of tourism, China has seen tremendous growth during the past thirty years, and it is predicted to become the most visited international destination by 2015–17 (Guangrui and Lew 2003; Wen and Tisdell 2001). Following the inception of the open-door policy of 1978, inbound tourism began to increase, and trade began to flourish. Since more openness in the early 1990s, tourism has grown dramatically, with many years in the 1990s and early 2000s seeing double-digit growth in international arrivals.

With increased trade and external contact, the standard of living of many Chinese improved; as a result, demand increased for travel experiences, and reluctantly the government began to allow Chinese citizens more freedom to travel within their own country (Guangrui 1989). This was particularly important, as many could not afford to travel abroad, and many desirable countries had not been granted the government's Approved Destination Status (ADS). Domestic tourism also achieved notable levels during the 1990s and early 2000s with between 6 and 11 percent annual growth. In 2005, at least 1.2 billion trips were taken by the Chinese within their own country (Lew *et al.* 2008). With the growth of inbound and domestic tourism, government policies underwent an evolution in the 1980s and 1990s from restraint to non-intervention and finally to positive encouragement (Wu *et al.* 2000: 298).

The same evolution occurred with outbound travel, albeit a bit later. Numbers of Chinese traveling abroad increased nearly fifteen times between 1994 and 2004, although the government still restricts the countries to which Chinese pleasure tourists may travel (Guo *et al.* 2007). Some ninety countries in Asia, the Pacific, Europe, Africa, and Latin America have been approved by the Chinese government as leisure tourism destinations for Chinese citizens. The United States received ADS in December 2007, and Canada is expected to be added to the list in the near future; those that have been added are seeing significant growth in Chinese arrivals (Chow and Murphy 2007; Guo *et al.* 2007; Pan and Laws 2003).

As a destination, China is best known for its cultural heritage. This does not, however, discount the importance of natural attractions as a focus of tourist attention. Because China is large and so culturally diverse, this chapter only highlights a few of the issues and trends taking place there. Many others exist but, because of space constraints, not all can be considered.

World Heritage Sites

China is home to thirty-seven UNESCO World Heritage Sites (WHS) (as of July 2008) (Figure 6.1), the third highest number, after Italy and Spain. Not unlike administrators in other less-developed countries, Chinese tourism officials are under a misguided impression that, once a heritage attraction is pronounced a World Heritage Site, the site will inevitably be deluged with international visitors and the area's economic woes will be a thing of the past (Su and Teo 2008; Yan and Morrison 2007). Unfortunately, this erroneous assumption has resulted in a lack of attention to content and quality at the country's heritage sites (Yan and Morrison 2007).

Another important problem facing China's WHS is overcrowding, even as administrators want more visitors. Crowding at heritage sites is not unique to China. It happens the world over, but it seems especially acute in China (du Cros 2007; Li *et al.* 2008), deteriorating both tourists' experiences and the physical environment. Many carrying capacities have been exceeded without adequate legislative protection and enforcement of extant legislation (du Cros 2007; Hall 1994).

Particularly notable as tourism continues to grow, based largely on WHS locations, is that local people have begun to lose control of their economies, cultures, and lifestyles. Perhaps one of the best documented examples today is Lijiang, a beautiful ancient city that was listed by UNESCO in 1997. In Lijiang, many of the indigenous Naxi people have been crowded out by more than three million visitors each year and the resultant unaffordable cost of living (Su and Teo 2008; du Cros 2007). These negative conditions are expected to continue with the recent implementation of Golden Weeks (additional Chinese holidays) and as international tourism continues to grow. As the Naxi move out, more people from other parts of the country are moving in to develop tourism services and retail establishments, people who are completely dissociated from, and unfamiliar with, the local culture. As Yamamura (2005: 198) notes:

> Naxi residents, who should arguably be the beneficiaries of a culture that is their own, are relatively weak in a competitive market ... the local Naxis, even if they want to enter the tourist trade, tend to miss opportunities ... it is relatively difficult for the local Naxis to compete with immigrant Han people in terms of capital and know-how.

In the case of Lijiang, this has resulted in a mass emigration with the exception of those who want to work directly in tourism and the elderly, who do not want to leave their homes (du Cros 2006: 209). This problem is not limited to Lijiang but is seen throughout China as more and more WHS are designated by UNESCO and as tourism continues to grow (Su and Teo 2008).

Nonetheless, while China's form of socialism in the past has essentially precluded the involvement and empowerment of local people in decision-making for tourism, the situation has begun to change. There is apparently a keener understanding among leaders that destination communities must be permitted to benefit from tourism, not just bear the brunt of its costs. Some heritage places, such as the Mount Huangshan WHS, are good examples in China of community-based development in recent years (Hatton 1999; Timothy and Tosun 2003).

Illegal trade in antiquities

One austere concern that has plagued China for many years, as well as many other developing countries, is the illicit trade in antiquities. Despite laws protecting historic artifacts, illegal digging and grave-robbing has existed for decades in some of China's most important cultural areas and continues to run rampant today, fueled by international collectors' markets in North America, Asia, and Europe, tourists purchasing antiquities and archeological relics from vendors, and poor people's need to survive (Atwood 2004; He 2001; Timothy and Boyd 2003). Unfortunately, this trend has resulted in the loss of many valuable artifacts and archeological sites throughout the country.

Ethnic minorities

Under Mao, China's policy toward its ethnic minorities was one of assimilation, where the sinicization of all ethnic groups was coerced in various ways to incorporate everyone into the mainstream Han culture. According to Bruner (2005), however, China's post-Mao ethnic policy has shifted from one of assimilation to one of acceptance and diversification. During the late 1970s and 1980s, in an effort to restore national unity after the trauma of the Cultural Revolution and to invigorate the economy through tourism, the national government began to accept ethnic minority cultures (fifty-five of them) as tolerable contributions to tourism and Chinese society at large (Sofield and Li 2007: 270). The development of cultural/ethnic parks highlights this policy change toward minority groups.

With the growth of domestic tourism in recent years and an increasingly affluent citizenry who are interested in things beyond their normal environments, demand for cultural experiences within China has increased. In response, during the post-Mao era, "ethnic parks" began to be developed by the government or private investors in various parts of the country to cater to the growing demand for domestic leisure travel (Li 2008; McKercher and du Cros 2002; Swain 1989; Wu et al. 2000; Zhong et al. 2006) and to function as "vehicle[s] for nation building" (Bruner 2005: 212). To bring China's cultures to the masses, these attractions tend to be situated near large population centers and attract primarily Mainland Chinese (80 percent), with visitors from other areas (including Hong Kong, Macao, and Taiwan) comprising the remaining 20 percent (Bruner 2005). Among the most impressive are Splendid China, Window of the World, and the Chinese Folk Cultural Village in Shenzen (Wu et al. 2000). Other ethnic parks and villages are located in the heart of ethnic regions, such as the National Minorities Park in Xishuangbanna, Yunnan Province—a popular destination for Chinese nationals which, according to Bruner's (2005: 217) tour guide, is not suited for foreign tourists because it is "too crowded with Chinese tourists, feature[s] fake reconstructed villages, and [is] not real." Unfortunately, among domestic travelers, particularly men, these regional folk villages and ethnic parks are often seen as places where non-Han minority women are:

> feminized exotic Others who ... are sexually promiscuous, erotically titillating, and available ... This gendered image of the minority female body is reflected at the Xishuangbanna theme park in many ways: the ethnic

houses feature attractive young women who are openly flirtatious, women of the Hani minority groups give massages, a transvestite performs at a show of minority dances, sex tours to Burma are available as a one-day side trip, and there are mock marriages and sexual banter between performers and tourists. ... and some of the performers are also sex workers.

Bruner (2005: 215)

The Chinese diaspora

One of the most important forms of cultural heritage tourism is that of diaspora-related travel, where immigrant populations and their descendants travel to the lands of their forebears, to discover their roots, to seek their own identity, or to visit distant relatives. This form of personal heritage travel is very salient in China, because of its diasporic connections throughout the world. Some recent estimates place the number of Overseas Chinese (people of Chinese origin) at well over 40 million, with especially large numbers in Asia, North America, and Europe (Tan *et al.* 2007; World Business 2007). Many of these people are still closely connected to their ancestral homeland and travel to China to visit ancestral villages, trace their roots, practice speaking Chinese, visit heritage and natural sites, and simply to be immersed in the land of their ancestors (Lew and Wong 2004).

As part of this broader personal heritage phenomenon, diasporic youth groups regularly travel to China in a "homecoming" heritage program sponsored by various philanthropic organizations to familiarize Overseas Chinese youth with their familial homeland. The Chinese Youth League of Australia and the China Youth Travel Service are two such organizations that sponsor ethnically based tours for Chinese young people from around the world who visit places in the country that best represent Chinese identity (Louie 2003).

Heritage and tourism in the Democratic People's Republic of Korea

The Korean Peninsula is divided into North Korea and South Korea by a 1953 ceasefire line known as the Demilitarized Zone (DMZ). Following the establishment of this border, the North (the Democratic People's Republic of Korea (DPRK)) and the South (The Republic of Korea (ROK)) have grown socio-economically in very divergent ways. South Korea has become a thriving capitalist and democratic country, while the leaders of North Korea elected to rule their nation with a unique form of communism known as *Juche*, which is a mix of Stalinism–Marxism and contemporary philosophies and directives of the late leader, Kim Il-Sung and his son, the current president, Kim Jeong-Il (Kim *et al.* 2007). The *Juche* system dictates every aspect of daily life in North Korea and keeps the citizenry devoted to its leaders. It has also slowed economic development owing to its focus on self-sufficiency to the full exclusion of outside control and influence, including trade and tourism, until fairly recently (Cho 2003; Hall 1990, 2001). In recent years, this has resulted in widespread starvation, energy shortages, and intervention by the world community to assist the isolated country in allowing more cross-border trade with Japan and other countries.

Since its inception in 1948, the DPRK has severely restricted travel to, from, and within its territory. Only recently has it begun to allow foreigners to travel there on guided group tours. Japanese, European, and Australian visitors began visiting in significant numbers in the 1980s and early 1990s; Americans were and continue to be prohibited from visiting the North, although a few select visas have been issued since 2002. More recently (late 1990s), owing to an economic arrangement between Hyundai Asan Corporation of South Korea and the DPRK government, and based on the North's need for foreign exchange, the DPRK began allowing South Koreans to visit the Mount Gumgang area on the east coast near the DMZ—a beautiful mountain area known to be sacred to all Koreans (Cho 2007; Kim and Prideaux 2003). A railway link opened in 2007 for limited cross-border trade, and tourists can now cross the border by coach as well. Some observers have suggested that these changes denote a linking together of the heritages of north and south and might eventually result in reunification or at least cordial relations between two countries that have until recently never communicated or cooperated (Kim and Prideaux 2003; Kim et al. 2006; Shin 2005).

Juche—*communism Korean style*

Essentially all tourism in North Korea can be classified as heritage tourism. From a heritage perspective, all tourism in the country is political in nature, and the two most significant matters in North Korean tourism are the Korean War/DMZ and the extreme socialist politicization of the past. In common with other socialist states, tourism in DPRK is highly political, perhaps more so than in any other country. Of the political uses of tourism outlined by Kim et al. (2007), the most salient for North Korea are the use of tourism as a tool to spread the *Juche* ideology to foreign visitors and to build patriotism and obedience among its own people. In DPRK, tourism is all about heritage— the *Juche*/communist heritage and ideals of the government. With the exception of tours to Mount Gumgang, which focus on nature, organized tours of the North still focus on visits to schools to highlight the high educational standards of the state, statues and monuments to the great leaders, other sites associated with the glories of the state (i.e., national hero tombs), and the DMZ to explain their version of the Korean War and the aggressors (South Korea and the United States). North Koreans themselves are unable to travel abroad, and their domestic travels are strictly controlled as well. In most cases, aside from visiting relatives, domestic travel in DPRK is sponsored and organized by schools and workplaces, and it spotlights studying the *Juche* dogma and visiting places associated with national heroes and the great leaders (Kim et al. 2007). National heritage is thus used as a propaganda tool for both national and international travelers.

The DMZ

The second main element of heritage tourism in North Korea is the DMZ and the Korean War (Henderson 2002). Several war memorials in the South attract Korean domestic tourists and Korean War veterans from the United States. Similar monuments exist in the North to declare its victory over the United States and ROK in 1953, and such monuments feature prominently in organized tours of that country. However, the most intriguing constituent of war tourism in the North (and in the South) is the DMZ—the border between North and South Korea (Lew *et al.* 2008). On the southern side of the line, daily tours visit several locations on the border, including Panmunjom, the truce village that lies within the Zone. Tours in the North also visit Panmunjom to demonstrate the benevolence of the *Juche* system and DPRK's willingness to cooperate with the outside world. The stories told on each side are quite opposite, however, each side claiming victory and labeling the other side the aggressor (Timothy *et al.* 2004).

There is an upsurge in border-related tourism throughout the world, where borders of conflict are becoming important heritage attractions (Gelbman 2008; Timothy 2001), falling somewhere within the overlapping domains of battlefield/war heritage, dark tourism (thanatourism), and political tourism. Panmunjom and the DMZ are one of the best and most unique examples of this, where soldiers serve as tourist guides, museums explain the Korean War from each vantage point, and lookout towers allow South Koreans and foreign tourists to look into the forbidden North and North Koreans to gaze into the ever-threatening South (Cho 2007; Lee 2006; Shin 2005, 2006; Timothy *et al.* 2004).

Mongolia's heritage tourism sector

As already noted, Mongolia is at a critical crossroads in terms of tourism. It is in the process of developing beyond the Soviet-style communist system that ruled the country until the peaceful democratic revolution of 1990. Since the establishment of a capitalist–democratic socio-political system, the country's literacy rate and life expectancy have increased. Likewise, in terms of international trade, commerce, and tourism, Mongolia is beginning to demonstrate a global presence and has seen considerable economic growth, although one-third of the country's population still lives in poverty (National Statistical Office 2007).

Like the countries of Eastern Europe and other former socialist states, Mongolia has experienced significant changes in the realm of tourism. Under the communist regime, international tourism to and from Mongolia was firmly controlled. Most international arrivals were from the Soviet Union and other countries of the Eastern Bloc and, when state-socialism collapsed in Europe, travel to Mongolia decreased dramatically. However, the country began to stabilize in 1998 when international arrivals increased by two-thirds over 1997 (Yu and Goulden 2006). Most years since then have seen salient growth in international arrivals, even in 2001 when many other destinations suffered losses; the year between 2001 and 2002 witnessed a jump of 20 percent (Gansukh and Tsermaa 2003). In 2004, Mongolia's income from tourism was estimated to be some US$181 million, comprising 10 percent of the country's GDP (Buckley *et al.* 2008). Tourism is now considered a critical part of an economy that has traditionally been based on mining and nomadic agriculture.

The cultural landscape and nomadism

Mongolians were, and many continue to be, a nomadic people, scattered in small clan groups throughout a sparsely populated desert and steppe terrain. Life itself was traditionally based on herding five kinds of animals: cattle and yaks, camels, horses, goats, and sheep. Products from these animals were the foundations of survival, and all of a family's needs were generally provided by such livestock. Arguably the most salient and recognizable material object in Mongolian culture is the ger (yurt)—the portable, circular homes made from sheep wool and camel hair. Today, more than half of Mongolia's population lives in gers on the steppes or as fixed residences in cities and towns (Thrift 2001).

Because Mongolians are a nomadic people, scattered throughout a sparsely populated terrain with only a few minor villages and towns, there is not a strong presence of built heritage or tangible artifacts to form the basis of heritage tourism. In some regards, this is a significant challenge for tourism developers. In spite of the dearth of built heritage, Mongolia has a strong oral culture, interesting customs and religious beliefs, and an avid interest in traditional sport (Thrift 2001), as well as a system of impressive national parks that are home to both natural and cultural elements (Bedunah and Schmidt 2000; Saffery 1999).

The most recognized element of Mongolian cultural heritage and the primary draw for foreign tourists is the marriage of its natural environment (steppes and deserts) and living culture. Mongolia is one of the most pronounced places in the world where natural landscapes and the cultural imprint on them are a true and completely interconnected heritage landscape, recognized as such by the world community. As noted above, because Mongols have traditionally been nomadic and rural, little by way of built heritage has been left for present-day use. In fact, it is the very intangible idea of the landscapes of human–nature interdependence that forms the core of the heritage product of Mongolia and features most prominently in its promotional endeavors (Buckley *et al.* 2008; Lew *et al.* 2008; O'Gorman and Thompson 2007). In essence, the entire country has been mythologized into a romantic rural landscape (by outsiders and tourism promoters), where small groups of nomads live in gers, herd horses, camels, and yaks, and subsist entirely off the land (Gansukh and Tsermaa 2003; Luvsandavaajav 2006; Yu and Goulden 2006) and where visitors " ... are wise to get out of the city, to tour with the nomads, whose territory begins about an inch beyond the last city building" (Horgan 2005: T7).

Nomadic life also gave rise to horse-based sport and various forms of gaming. Sports and games practiced centuries ago are still practiced in the countryside today, and every year, tens of thousands of people from all over the country and from abroad congregate in Ulaanbaatar to participate in and witness the Naadam Festival. The festival focuses on nomadic hunting and warring traditions, namely wrestling, archery, and horse racing, and highlights celebrations of gratitude for health and wealth (Schofield and Thompson 2007). Naadam is a significant nationalistic event for Mongolians and an original and "authentic" Mongolian experience for foreign spectators (O'Gorman and Thompson 2007; Schofield and Thompson 2007).

Tangible heritage

While the primary focus inside and outside Mongolia is the country's intangible, living heritage, there is also a built presence that forms an important element of the heritage milieu, particularly in towns and cities expanded by the Soviets between the 1920s and the 1990s. Primary among these are the Winter Palace of Bogd Khan; the ancient capital, Kharakhorum; and various monasteries in Ulaanbaatar and other regional centers (Thrift 2001; Warner 2002).

Mongolia is continuing to break from its communist past, even though several communist leaders were elected to run the country following open elections in the early 1990s. In countries such as Hungary, Romania, and Bulgaria, populations are still coming to terms with their socialist past and have reacted to it in a variety of ways, including avoidance, destruction, reminiscing, and conservation. Unlike some countries of the former Soviet Union and Eastern Europe, Mongolia does not have the same difficulties. There are few tangible remains in Mongolia today to testify to the country's communist past, with the exception of Soviet-era block buildings incongruously abutting traditional gers in urban areas. Likewise, there are still a few signs in the urban landscape of Ulaanbaatar that bring to mind the memory of the political past, such as statues of Lenin and the Zaisan Monument, which memorializes Soviet soldiers killed in the Second World War. Unfortunately, however, one legacy of the communist regime was the destruction of nearly 800 monasteries throughout the country in 1937 that were an integral part of Mongolian Buddhist culture and religion.

Challenges

Despite its rich cultural traditions and romanticized cultural landscapes, everything is not rosy in Mongolian tourism (Luvsandavaajav 2006). Like other developing regions, Mongolia faces budget and staffing shortfalls that prevent adequate conservation of the built environment and interpretation of the country's intangible heritage. Perhaps more vexing to the industry and tourists is the lack of accessibility to some of the most scenic and important heritage areas. Roads and highways are inadequate, as many are unsurfaced, and travel times are long and arduous. This often limits the range of places a tourist on a timetable can visit. Likewise, the country is well connected with air hubs to and from Ulaanbaatar, but there are few flights between regional centers; essentially all domestic flights must travel back and forth through the capital. Finally, while tourist arrivals are increasing, the tourism infrastructure is not keeping up. There is, for instance, an inadequate supply of accommodations outside of Ulaanbaatar to meet the needs of tourists.

Discussion and conclusion

The developing parts of East Asia have great potential for tourism growth, and most have seen increased arrivals in recent years. Facing all three countries, particularly in heritage site management, is a lack of planning and tourism management authorities to deal with specific sites and address each one's unique challenges and opportunities. Not surprisingly, a lack of funding also creates problems related to conservation, training, and interpretation. The developing countries of East Asia are characterized by a communist–socialist past that heavily influences each nation's heritage to some degree or another in terms of supply, demand, interpretation, utilization, and even management and planning. However, each of the three countries demonstrates different levels of what is expected in heritage tourism in state-socialist systems. North Korea still epitomizes the historical pattern of heritage tourism in socialist states, wherein the past is used openly as a political tool. China, while still a communist state, demonstrates few if any of the traditional elements of socialism-based tourism, and Mongolia appears to have forgone its communist past almost entirely, with the exception of modern architecture and a few remaining landscape features that resemble the days of Soviet domination.

This socialist legacy overshadows a large share of the management and planning challenges facing all three countries. All less-developed countries confront problems regarding empowerment of people and participatory development. However, for those that have experienced a socialist past, there is an added dimension—a strict form of top-down planning that essentially disallowed all forms of participation in tourism, sometimes even precluding people from working in the industry if they so desired. Although Mongolia and China have overcome many of the challenges posed by this socio-political legacy, they still face many issues of interpretation, conservation, and participatory management and planning. The existing system in North Korea is the most concerning issue facing the heritage of that country. Much of the past there has been replaced by a new past told by the Great Leaders to develop loyalty among the country's citizenry (French 2007; Oh and Hassig 1999).

Despite these challenges, there are more positives than negatives. At present, all three countries commonly exude a sense of mysticism in regards to heritage, which makes them desirable tourist destinations. China is home to mysterious and unique cultural groups, as well as heritage sites that are world-renowned symbols of the past (e.g., the Great Wall and the Terracotta Warriors). North Korea's strict closed-door policy, particularly for some nationalities, creates a mystical sense of "the unknown," which automatically places it on some people's "must-do" list. Mongolia's cultural landscapes, which have been well maintained through the centuries, emanate an extraordinary appeal for outsiders as a surreal place only dreamt about, read about in books and magazines, or visualized on TV and in movies. The issues, attractions, challenges, and opportunities highlighted in this chapter only scratch the surface of a region that is rich in many pasts that tourists desire to experience.

References

Atwood, R. (2004) *Stealing History: Tomb Raiders, Smugglers, and the Looting of the Ancient World*. New York: St. Martin's Press.

Bedunah, D.J. and Schmidt, S.M. (2000) Rangelands of Gobi Gurvan Saikhan National Conservation Park, Mongolia. *Rangelands*, 22(4): 18–24.

Bruner, E.M. (2005) *Culture on Tour: Ethnographies of Travel*. Chicago, IL: University of Chicago Press.

Buckley, R., Ollenburg, C. and Zhong, L. (2008) Cultural landscape in Mongolian tourism. *Annals of Tourism Research*, 35(1): 47–61.

Chang, K.C. (1991) Ancient China and its anthropological significance. In C.C. Lamberg-Karlovsky (ed.), *Archaeological Thought in America*, pp. 155–66. Cambridge: Cambridge University Press.

Cho, M. (2007) A re-examination of tourism and peace: the case of the Mt. Gumgang tourism development on the Korean Peninsula. *Tourism Management*, 28(2): 556–69.

Cho, M.C. (2003) Current status of the North Korean economy. In C.Y. Ahn (ed.), *North Korea Development Report 2002/2003*, pp. 32–51. Seoul: Korea Institute for International Economic Policy.

Chow, I. and Murphy, P. (2007) Travel activity preferences of Chinese outbound tourists for overseas destinations. *Journal of Hospitality and Leisure Marketing*, 16(1/2): 61–80.

du Cros, H. (2006) Managing visitor impacts at Lijian, China. In A. Leask and A. Fyall (eds), *Managing World Heritage Sites*, pp. 205–14. Amsterdam: Elsevier.

—— (2007) Too much of a good thing? Visitor congestion management issues for popular World Heritage tourist attractions. *Journal of Heritage Tourism*, 2(3): 225–38.

French, P. (2007) *North Korea: the Paranoid Peninsula—A Modern History*. London: Zed Books.

Gansukh, D. and Tsermaa, B. (2003) Rural tourism in Mongolia. In J.K. Jun (ed.), *Proceedings of the Second Asia Pacific Forum for Graduate Student Research in Tourism*, pp. 646–49. Pusan: Dong-A University and the Korea Academic Society of Tourism and Leisure.

Gelbman, A. (2008) Border tourism in Israel: conflict, peace, fear and hope. *Tourism Geographies*, 10(2): 193–213.

Gernet, J. (1996) *A History of Chinese Civilization*. Cambridge: Cambridge University Press.

Guangrui, Z. (1989) Ten years of Chinese tourism: profile and assessment. *Tourism Management*, 19(1): 51–62.

Guangrui, Z. and Lew, A.A. (2003) Introduction: China's tourism boom. In A.A. Lew, L. Yu, J. Ap and Z. Guangrui (eds), *Tourism in China*, pp. 3–12. New York: Haworth.

Guo, Y., Kim, S.S. and Timothy, D.J. (2007) Development characteristics and implications of Mainland Chinese outbound tourism. *Asia Pacific Journal of Tourism Research*, 12(4): 313–32.

Hall, C.M. (1994) *Tourism in the Pacific Rim: Development, Impacts and Markets.* Melbourne: Longman.

Hall, D.R. (1990) Stalinism and tourism: a study of Albania and North Korea. *Annals of Tourism Research*, 17: 36–54.

—— (2001) Tourism and development in communist and post-communist societies. In D. Harrison (ed.), *Tourism and the Less Developed World: Issues and Case Studies*, pp. 91–107. Wallingford: CABI.

Hatton, M.J. (1999) *Community-Based Tourism in the Asia-Pacific*. Ottawa: Canadian International Development Agency.

He, S. (2001) Illicit excavation in contemporary China. In N. Brodie, J. Doole and C. Renfrew (eds), *Trade in Illicit Antiquities: The Destruction of the World's Archaeological Heritage*, pp. 19–24. Cambridge: McDonald Institute.

Henderson, J.C. (2002) Tourism and politics in the Korean Peninsula. *Journal of Tourism Studies*, 13(2): 16–27.

Horgan, D. (2005) Steppe forward: Mongolia slowly opening to tourism. *Arizona Republic*, October 2: T7.

Kim, S.S. and Prideaux, B. (2003) Tourism, peace, politics and ideology: impacts of the Mt. Gumgang tour project in the Korean Peninsula. *Tourism Management*, 24 (6): 675–85.

Kim, S.S., Lee, H. and Timothy, D.J. (2006) Perspectives on inter-Korean cooperation in tourism. *Tourism Analysis*, 11(1): 13–23.

Kim, S.S., Timothy, D.J. and Han, H.C. (2007) Tourism and political ideologies: a case of tourism in North Korea. *Tourism Management*, 28(4): 1031–43.

Lee, Y.S. (2006) The Korean War and tourism: legacy of the war on the development of the tourism industry in South Korea. *International Journal of Tourism Research*, 8: 157–70.

Lew, A.A. and Wong, A. (2004) Sojourners, *guanxi* and clan associations: social capital and overseas Chinese tourism to China. In T. Coles and D.J. Timothy (eds), *Tourism, Diasporas and Space*, pp. 202–14. London: Routledge.

Lew, A.A., Hall, C.M. and Timothy, D.J. (2008) *World Geography of Travel and Tourism: A Regional Approach*. Oxford: Butterworth Heinemann.

Li, J. (2008) Tourism enterprises, the state, and the construction of multiple Dai cultures in contemporary Xishuang Banna, China. In B. Prideaux, D.J. Timothy and K.S. Chon (eds), *Cultural and Heritage Tourism in Asia and the Pacific*, pp. 205–20. London: Routledge.

Li, M., Wu, B., and Cai, L. (2008) Tourism development of World Heritage Sites in China: a geographic perspective. *Tourism Management*, 29(2): 308–19.

Louie, A. (2003) When you are related to the "other": (re)locating the Chinese homeland in Asian American politics through cultural tourism. *Positions*, 11(3): 735–63.

Luvsandavaajav, O. (2006) Mongolia's image as a tourist destination: perceptions of foreign tourists. Paper presented at the 3rd Graduate Research in Tourism Conference, Ankara, Turkey, May 25–28.

McKercher, B. and du Cros, H. (2002) *Cultural Tourism: The Partnership Between Tourism and Cultural Heritage Management*. New York: Haworth.

National Statistical Office (2007) *Statistical Yearbook of Mongolia, 2006*. Ulaanbaatar: National Statistical Office.

O'Gorman, K. and Thompson, K. (2007) Tourism and culture in Mongolia: the case of the Ulaanbaatar Nadaam. In R. Butler and T. Hinch (eds), *Tourism and Indigenous Peoples: Issues and Implications*, pp. 161–75. Oxford: Butterworth Heinemann.

Oh, K. and Hassig, R. (1999) North Korea between collapse and reform. *Asian Survey*, 39(2): 287–309.

Pan, G.W. and Laws, E. (2003) Tourism development of Australia as a sustained preferred destination for Chinese tourists. *Asia Pacific Journal of Tourism Research*, 8 (1): 37–47.

Saffery, A. (1999) Mongolia's tourism development race: case study from the Gobi Gurvansaikhan National Park. In P.M. Godde, M.F. Price and F.M. Zimmermann (eds), *Tourism and Development in Mountain Regions*, pp. 255–74. Wallingford: CABI.

Schofield, P. and Thompson, K. (2007) Visitor motivation, satisfaction and behavioural intention: the 2005 Naadam Festival, Ulaanbaatar. *International Journal of Tourism Research*, 9: 329–44.

Shin, Y.S. (2005) Safety, security and peace tourism: the case of the DMZ area. *Asia Pacific Journal of Tourism Research*, 10(4): 411–26.

—— (2006) Perception differences between domestic and international visitors in the tourist destination: the case of the borderline, the DMZ area. *Journal of Travel and Tourism Marketing*, 21(2/3): 77–88.

Sofield, T. and Li, F.M. (2007) Indigenous minorities of China and effects of tourism. In R. Butler and T. Hinch (eds), *Tourism and Indigenous Peoples: Issues and Implications*, pp. 265–80. Oxford: Butterworth Heinemann.

Su, X.B. and Teo, P. (2008) Tourism politics in Lijiang, China: an analysis of state and local interactions in tourism development. *Tourism Geographies*, 10(2): 150–68.

Swain, M.B. (1989) Gender roles in indigenous tourism: Kuna Mola, Kuna Yala, and cultural survival. In V. Smith (ed.), *Hosts and Guests: The Anthropology of Tourism*, pp. 83–104. Philadelphia, PA: University of Pennsylvania Press.

Tan, C.B., Storey, C. and Zimmerman, J. (2007) *Chinese Overseas: Migration, Research and Documentation*. Hong Kong: Chinese University Press.

Thrift, E. (2001) *The Cultural Heritage of Mongolia*. Ulaanbaatar: Naranbulag Publishing.

Timothy, D.J. (2001) *Tourism and Political Boundaries*. London: Routledge.

Timothy, D.J. and Boyd, S.W. (2003) *Heritage Tourism*. Harlow: Prentice Hall.

Timothy, D.J. and Tosun, C. (2003) Appropriate planning for tourism in destination communities: participation, incremental growth and collaboration. In S. Singh, D.J. Timothy and R.K. Dowling (eds), *Tourism in Destination Communities*, pp. 181–204. Wallingford: CABI.

Timothy, D.J., Prideaux, B., and Kim, S.S. (2004) Tourism at borders of conflict and (de)militarized zones. In T.V. Singh (ed.), *New Horizons in Tourism: Strange Experiences and Stranger Practices*, pp. 83–94. Wallingford: CABI.

Warner, G.A. (2002) Mongolia restoring past, embracing future. *Arizona Republic*, November 3: T8.

Wen, J.J. and Tisdell, C.A. (2001) *Tourism and China's Development: Policies, Regional Economic Growth, and Ecotourism*. Singapore: World Scientific.

World Business (2007) The world's successful diasporas. World Business, April 3. Available from http://www.worldbusinesslive.com/research/article/648273/the-worlds-successful-diasporas/ (accessed July 1, 2008).

Wu, B., Zhu, H. and Xu, X. (2000) Trends in China's domestic tourism development at the turn of the century. *International Journal of Contemporary Hospitality Management*, 12(5): 296–99.

Yamamura, T. (2005) Dongba art in Lijian, China: indigenous culture, local community and tourism. In C. Ryan and M. Aicken (eds), *Indigenous Tourism: The Commodification and Management of Culture*, pp. 181–99. Amsterdam: Elsevier.

Yan, C. and Morrison, A.M. (2007) The influence of visitors' awareness of World Heritage listings: a case study of Huangshan, Xidi and Hongcun in Southern Anhui, China. *Journal of Heritage Tourism*, 2(3): 184–95.

Yu, L. and Goulden, M. (2006) A comparative analysis of international tourists' satisfaction in Mongolia. *Tourism Management*, 27(6): 1331–42.

Zhong, J., Chen, B. and Yang, G.H. (2006) The tourist experience levels of the ethnic villages in Yunnan. *China Tourism Research*, 2(1/2): 71–76.

7 Heritage tourism in the Pacific

Modernity, myth, and identity

C. Michael Hall

Introduction

In the minds of many Western tourists, the idea of the Pacific conjures up impressions of swaying tropical palm trees, white sand beaches, warm crystal-clear waters and, possibly, dusky maidens in grass skirts or sarongs. This stereotypical, and highly gendered, image of "paradise" has been consistently portrayed over many years, not only in tourist advertising, but also in many other forms of image making, such as film, newspapers and magazines, novels, music, and even academic works (Connell and Gibson 2008; Douglas 1996; Harrison 2003, 2004; Sturma 2002). What is remarkable about this image is its consistency for much of the past 200 years and probably longer (Hall 1998). Nevertheless, as Harrison (2004: 2) noted, "like all stereotypes, that of the tropical island paradise contains some truth but much inaccuracy."

Image making is essential to tourism. Tourism, perhaps more than any other business, is based on the production, reproduction, and reinforcement of images. These images serve to project the "other" into the lives of consumers and, if successful, will assist in setting the socially constructed boundaries of a network of attractions, which is referred to as "a destination" (Hall 1998).

Otherness is a significant component in tourism marketing and the establishment of cultural and heritage stereotypes. "Encounters with the 'other' have always provided fuel for myths and mythical language. Contemporary tourism has developed its own promotional lexicon and repertoire of myths ... " (Selwyn 1993: 136). For many visitors, otherness is what makes a destination worthy of consumption. Although, ironically, as Hitchcock *et al.* (1993: 3) observed, "large numbers of tourists may be attracted to the region by its perceived 'differentness', lured by the images of culture and landscape which are vividly portrayed in the promotional literature, few are able or willing to tolerate a great deal of novelty." However, to build binary opposites is to make one dependent on the other. There cannot be consumption without production. "It is apparent that they merge in many places and that each process certainly does have effects on the other ... even if they are causal or may never ever be explicable" (Laurier 1993: 272). Any understanding of the creation of a destination therefore involves placing the development of the representation of that destination within the context of the consumption and production of places and, more particularly, the manner by which places have become incorporated within the global system that provides not only for economic exchange but also the commodification of culture and heritage as a means for facilitating the accumulation of capital within the system. The emergence of capitalism in Europe in the Middle Ages coincided with the imperialist ambitions of the European powers, improvements in transport technology, and the development of a mercantile class that sought raw resources, produce, and trade with the ever-widening expanse of the European known world. Exploration became a geographical activity driven by the urgencies of economic growth. As Hall (1998: 141) argued:

> The Pacific as a destination is a creation of capitalism. While the Pacific Ocean serves to provide a physical boundary for the Pacific, it is the socially constructed Pacific with its attendant myths which dominates the tourist consumers mind and is commodified for the tourist's pleasure (and capital) which is by far the most important. However, the place making of the Pacific is not just a tourism phenomenon, it must be related to the means by which the Pacific was incorporated into the global capitalist system.

Such issues are important for heritage tourism as conceptualizations of heritage are constructed both within a culture and from the outside via the weight of external market expectations and understandings of what constitutes heritage for a given location or community. In the case of the Pacific, which has particular colonial, post-colonial, and neo-colonial histories, there are therefore different fragments and strands of heritage and identity, some of which come from the people of the Pacific themselves but which is also derived from external framing of Pacific cultures, initially through the colonial mercantile economy but, more recently, via the global tourist economy.

This chapter aims to draw together a number of these different dimensions of heritage in relation to heritage tourism practices. This is done initially via describing the extent by which the cultures of the Pacific islands have and continue to be marketed in Romantic terms and then in terms of the creation of different elements of heritage tourism. However, before examining the imaging of the Pacific islands, the chapter will briefly turn to the economic importance of tourism to the region.

Tourism in the Pacific

The Pacific island nations have very few economic resources. Because of their colonial history, small size, and distance from major markets, they have an extremely small indigenous capital base. They are therefore reliant on foreign powers to provide capital for economic development and the transport links that enable the export of goods and services. At the beginning of the twenty-first century, it is the countries of the Pacific Rim that have made great strides in economic development, not the island states of the Pacific Ocean. The problems facing the island nations of the Pacific are typical of those that confront nearly all of the world's small island nations or island microstates. They lie at the margins of the global economy, are highly dependent on foreign aid and investment programs, as well as financial remittances (Browne 2006; Browne and Mineshima 2007). Furthermore, they have relatively little control over their scarce natural resources, and have relatively little power to influence the economic and political direction of the region in which they are situated (Hall and Page 1996; Harrison 2004).

Pacific island economies generally share a common feature in that they are net importers with minimal capacity to generate foreign exchange independently (Department of Foreign Affairs and Trade 1994). The Pacific islands have few natural resources that can be exploited and those that do exist, such as fish, minerals, and timber, are under pressure of over-exploitation and the lack of economic alternatives and development options. Furthermore, many of the economies of the Pacific islands are based on one or two commodities that are subject to significant price fluctuations. "The outcome is that they generally have very few products to sell to sophisticated markets. There exists a small fragile private sector to pursue opportunities as they arise, but obtaining suitable, quality venture capital to finance sustainable globally competitive ventures is a chronic problem" (Department of Foreign Affairs and Trade 1994: np). It is therefore perhaps not surprising that, given the need to diversify their economic bases, rising social expectations, and increasing population pressures, great importance has been attached by Pacific island governments to the development of service industries, such as financial services and tourism, as a means of making an important contribution to economic growth and employment (Department of Foreign Affairs and Trade 1994; Hall 1997; Harrison 2004). Because of these circumstances, Connell's (1988: 62) observation that "For island states that have very few resources, virtually the only resources where there may be some comparative advantage in favour of [island microstates] are clean beaches, unpolluted seas and warm weather and water, and at least vestiges of distinctive cultures" holds as true today as it did when it was written.

Table 7.1 records some of the changes in visitor arrivals to the Pacific over time, as well as some key economic and demographic data. Figure 7.1 portrays the heritage sites in the Pacific that have been listed by UNESCO. With the exception of Fiji, Hawai'i, and French Polynesia, visitor growth continued in 2007 when visitors to the South Pacific grew by an overall 4.1 percent on 2006 figures to a sum of 1.335 million visitors. In commenting on the 2007 figures, south-pacific.travel (the virtual brand of the South Pacific Tourism Organisation (SPTO)) chief executive Tony Everitt said, "Given the increased cost of oil, this level of growth is a pleasing result. It shows that tourism continues to lead economic development in the South Pacific" (south-pacific. travel 2008: np).

Table 7.1 Tourism, economic, and demographic data for the Pacific (thousands), selected years, selected states

Year	Tourist arrivals 1990 (thousands)	Tourist arrivals 1995 (thousands)	Tourist arrivals 2000 (thousands)	Tourist arrivals 2005 (thousands)	Total expenditure of visitors 2005 (US$m)	Total GDP 2005 (US$m)	Service sector contribution to GDP 2005 (%)	Population 2006 (thousands)
Cook Islands	34	48	73	88	92	183	79.2	14
Fiji	279	318	294	550	575*	2,998	64.4	833
French Polynesia	132	172	252	208	767*	5,388	82.1	259
Kiribati	3	4	5	3*		72	78.2	94
Marshall Islands	5	6	5	9		111	70.9	58
Nauru						55	78.2	10
New Caledonia	87	86	110	101		4,341	72.9	238
Niue	1	2	2	3				2
Palau						123	83.3	20
Papua New Guinea	41	42	58	69		5,330	24.9	6,202
Samoa	48	68	88	102		406	59.8	185
Solomon Islands	9	12	6			299	45.4	484
Tonga	21	29	35	41*		214	56.9	100
Tuvalu	1	1	1	1*		26	69.8	
Vanuatu	35	44	58	61*		329	77.0	221
Total	696	862	987	1,236				
US territories								
American Samoa	26	49	44					65
Federated States of Micronesia		33				239		111
Guam	780	1,362	1,287	1,157*				
Hawai'i	6,971	6,629	6,949	7,417				171
Northern Mariana Islands	426	669	517	525*				

*2004 figures.
Source: UNCTAD (2007).

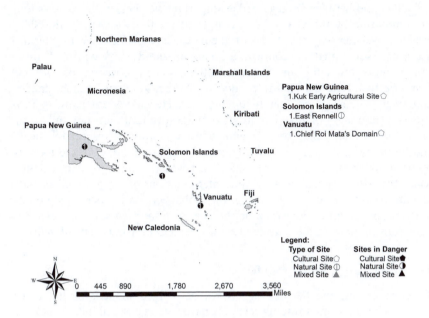

Figure 7.1a Developing Countries and World Heritage Sites in the Pacific

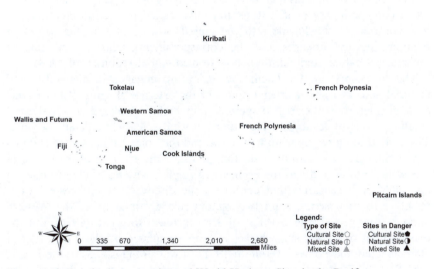

Figure 7.1b Developing countries and World Heritage Sites in the Pacific

Culture and heritage are regarded as important elements of the region's tourism promotion by the SPTO. For example, at the time of writing, the SPTO weekly newsletter featured stories on the perceived benefits of World Heritage listing for New Caledonia tourism, a Solomon Islands story on how "culture can promote tourism," and the international media coverage of the 2008 Tahiti Festival (Heiva I Tahiti), particularly "the traditional Tahitian dancing and singing competition that highlights each Heiva." Accompanying these stories were images of islanders in grass skirts (primarily female) and stories of the new Fijian China website and industry news with respect to "health and harmony in Paradise," which referred to health holidays at the Paradise resort in Fiji (south-pacific.travel 2008). The newsletter highlighted the extent to which the South Pacific relies on both interpretations of "traditional" islander identities as well as its integration in the global economic system to promote tourism opportunities. However, as the next section discusses, these elements have been in place in the outside view of the region for hundreds of years.

The exoticization of the Pacific

To some extent, the Pacific has always been a component of the global system. Migrations and trading relations have always ensured linkage between different peoples in the region and then to the wider world. However, in the pre-capitalist world, multiple centers of power and dominance existed, while the rate at which global trade and economic exchange and, arguably, cultural exchange was conducted was considerably slower when compared with today. The emergence of capitalism and the accompanying development of a particular set of core–periphery relationships and linkages led to the full creation and incorporation of the Pacific into the global system. This particular set of relationships and linkages and the corresponding increase in the rate of exchange of capital and culture is now termed globalization (Hall 1998).

The "discovery" of the Pacific by Europeans engaged in mercantile and imperial activity was the crucial point for the external imaging of the Pacific as a distinct romanticized, exoticized, and highly gendered other, thereby creating a weighty legacy of expectations of what is authentic Pacific heritage. The early trading relationship with India and the Spice Islands of the Indonesian archipelago was an initial starting point for the creation of the image of the exotic. However, it was the French and English voyages of the seventeenth and eighteenth centuries that confirmed the discovery of "paradise." Contributing to this picture were two factors strongly influencing the Western mind in this period: the writings of Jean-Jacques Rousseau (1978) and the reassessment of Classicism, which had been stimulated by the unearthing of Herculaneum and Pompeii (Honour 1981). It was in the islands of the Pacific that Rousseau's Romantic "noble savage," elements of which had already been identified in the peoples of the Americas and Southeast Asia, was to be discovered (Douglas 1996; Fulford and Lee 2002). When the French explorer Louis Antoine de Bougainville landed in Tahiti in April 1768,

He marvelled how Nature had endowed the island and allowed a people to live in happiness barely removed from the state of Nature. Sentiment rather than reason Vanquished unpleasant aspects of his stay, and in departing, stated his desire to forever "extol the happy isle of Cythera: it is the true Utopia."

Brown (1988: 12)

Bougainville was not alone in these sentiments. In the English-speaking world, publications of various accounts from Cook's voyages also served to establish the Romantic image of the Pacific in the European, and especially the British, imagination (Smith 1960), as well as profoundly influencing the notion of travel itself.

Joseph Banks applied the names of Greek heroes when naming the inhabitants of Tahiti. If Bougainville cherished visions of an "Island of Love," for Banks, Tahiti became the truest picture of "Arcadia" (Brown 1988: 12). Similarly, for the naturalist Georg Forster, who accompanied Cook on his second voyage, Tahiti provided, in his eyes, a setting for the unambitious living which knew neither the "absolute want nor the unbounded voluptuousness" of European society (Groves 1995: 317). Indeed, Forster was concerned as to the impact that European economic and colonial expansion would have on the Pacific Islanders, writing in 1777 that, "if the knowledge of a few individuals can only be acquired at such a price as the happiness of nations it were better for the discoverers and the discovered that the South Seas had remained unknown to Europe and its restless inhabitants" (cited in Groves 1995: 317).

The popular writing of the eighteenth and nineteenth centuries received reinforcement in the governmental images of the Pacific, which also promoted the Romantic (including the eroticization of islander women) and the picturesque. Arguably, this was done by "official media" for two major reasons. First, such an image was in keeping with the dominant intellectual fashion of the times. Second, images could be put to utilitarian ends. Government and commercial enterprises encouraged such images in order to encourage settlement and therefore to provide a firmer base for the incorporation of these new lands into the imperial structures and, consequently, into the global system (Hall 1998).

The idea of a Pacific Arcadia was therefore developed in great part to encourage a flow of migrants to the new worlds of the Pacific. So powerful was the initial promotion of this image of a better life in this world that it continues to this day, at least in the European and North American mind and in the promise of tourist advertising. The ability of tourism to evoke such an evocative image should come as no surprise. The promise of a better time is little different in either migration schemes or consumption of tourist packages. Moreover, the agencies responsible for migration and the encouragement of international tourism were often one and the same. In addition, the transport system that brought the migrants and facilitated international trade was the same system that served the early tourists. For example, Fiji has been a tourism destination since the early twentieth century when it was a regular stopping point for trans-Pacific shipping. The economic potential of tourism was officially recognized in 1924 when the Fiji Publicity Board was established to run a tourist bureau at the behest of the White Settlement League. The terms of reference for the Board were "to make recommendations with a view to popularizing the colony to tourists, to provide facilities to tourists to visit places of interest, and to consider the best suitable methods of providing funds for the objects it desired to attain" (Ministry of Tourism 1992: 1). Similarly, mercantile shipping connections between Hawai'i and the United States mainland served as the basis for both the annexation of the islands by the United States and the development of a tourism industry, to which commercial interests were applying the term "Paradise" by the 1850s (Douglas and Douglas 1996) and which later came to be used throughout the twentieth century with respect to a range of media representations of the islands and the wider Pacific (Bacchilega 2007; Connell and Gibson 2008).

The shipping network that overlay ancient migration and trading routes in the South Pacific was extremely important in creating a Pacific identity in the late nineteenth and early twentieth centuries. The network served to tie together the islands of the South Pacific with the metropolitan powers of Australia and New Zealand to the west and Canada and the United States to the east. Despite the vast distances and the different cultures and peoples in this region, Melanesia, Micronesia, and Polynesia, with Australia and New Zealand at the periphery, were drawn together into a common region in the eyes of the European and North American gaze. This is a regional and cultural identity that was only to receive reinforcement in the production of tourism images that are still with us to the present day.

Included in such images, and one of the paramount dimensions of island marketing and promotion, has been "the South Sea maiden." As Sturma (2002: 2) commented in his book on the subject:

> From early European contact in the Pacific, the South Sea maiden occupied a special place in the Western imagination. The "island girl" is an integral part of the adventure, sensuality, and romance associated with the South Pacific. She figures prominently in the writings of early Pacific voyagers, is central to a genre of fiction exemplified by writers from Melville to Michener, is discussed at length by anthropologists, is featured in numerous films, and is still a staple of tourism advertising. Although less in evidence these days than in the past, the South Sea maiden remains a powerful symbol.

The South Pacific maiden also became integral to the understanding and portrayal of the Pacific islands as a "paradise" by European male explorers. For example, in his written account of the *Bounty* mutiny, the ship's commander William Bligh emphasized the sexual dimension of the women of

Tahiti and their supposed role in tempting the ship's crew to mutiny, although as Sturma (2002: 37) notes, "Bligh, of course, had good reason to emphasize the seductive powers of the Tahitian women and distract attention from allegations that his tyranny drove the *Bounty* men to mutiny." Such literary accounts are important as they laid the foundations for the more recent portrayals of the Pacific as "paradise" or as "timeless," including numerous remakes of the *Mutiny on the Bounty*, on film and television. In fact, Sturma (2002: 137) observes that, "much of South Pacific tourist advertising is directed to maintaining the fiction that somehow in the process of becoming colonial outposts and tourist resorts, the islands have remained timeless and unchanged." Yet, tourism marketing, often in association with Western media representations of the islands, is vital to packaging "paradise."

For example, the Australian and New Zealand promotional campaign for Vanuatu in the 1990s used the theme of "Vanuatu the untouched paradise" and featured Australian musician's John Farnham's hit *Touch of Paradise*. According to the National Tourism Office of Vanuatu (1990), the campaign led to a rapid increase in visitors from Australia, which showed very healthy increases in the second half of 1989. However, as the *Pacific Islands Monthly* (1990: 38) observed, "skeptics may smile at both the originality and the accuracy of the slogan. Surely of all the Pacific's 'paradise,' Vanuatu has been touched more often than many? But the success of the campaign is beyond argument." Even in 2008, the website of Destination Vanuatu is http://www.vanuatuparadise.com/. Yet, as Douglas and Douglas (1996: 32–33) argued:

> The myth of Paradise is by now a thoroughly shop-worn cliché, which invests every kind of promotion ... Virtually every travel brochure on the region contains similar images, no longer the exclusive preserve of Tahiti, which inspired them, or Hawai'i which mass produced them. By the 1970s, aided by jet travel, packaged vacations and the relentlessness of brochure and television advertising, the myth had been exported more widely than any other regional product and was being applied indiscriminately and often incongruously to every part of the Pacific.

Indeed, they went on to note that, "The myth had become so pervasive that its presence was evident even in the work of those who ought to be critical of it" (Douglas and Douglas 1996: 34), and illustrate this by noting that Farrell, in his introduction to *Hawaii: The Legend That Sells*, is lured to its use thus: "Take a group of breathtakingly beautiful islands set in the blue Pacific as close to paradise as you wish ... " (Farrell 1982: xiii). Even the contemporary strategic tourism plan for Hawai'i still refers to "paradise":

> From the mountain to the ocean. Hawai'i offered, and continues to provide, stunning vistas, lush rainforests, dramatic mountain ranges, beautiful beaches, and a temperate climate – all conducive to creating a "paradise" in the middle of the Pacific. These first people, the native

Hawaiians, and Hawai'i's relative isolation have worked together to produce a distinctive destination found no where else in the world

State of Hawai'i (2005: 3)

Yet, all is not very well in paradise. In 2007, Hawai'i experienced a drop in tourism numbers for the first time since 2003, with a corresponding impact on employment and state tax revenue as a result of less visitor spending. Like many Pacific islands, tourism is economically extremely significant for Hawai'i, where it accounts for one-quarter of all spending (eTurboNews 2008). However, the state was one of the first Pacific destinations to be affected by the downturn in the US economy and reflects a 7.4 percent decline in US visitation to the South Pacific region in 2007 (Everitt 2008). The economic concerns we are exacerbated by the continuing high price of oil in 2008 and subsequent impacts on airlines servicing the region, while in the longer term, issues of climate and environmental change are also seen to be important to the many low-lying island groups in the Pacific (UNEP, UNWTO, and WMO 2008). Nevertheless, it is important not to repeat the romanticization of the past when assessing cultural and environmental change in the Pacific. As Sahlins (2005) notes with respect to development processes in the Pacific:

> When Europeans change it is called "progress", but when "they" (the others) change, notably when they adopt some of our progressive attributes, it is a loss of their culture, some kind of adulteration. As the European folklore goes, before we came upon the inhabitants of the Americas, Asia, Australia or the Pacific Islands, they were "pristine" and "aboriginal". It is as if they had no historical relations with other societies, were never forced to adapt their existence. Rather, until Europeans appeared, they were "isolated" – which means that we were not there. They were "remote" and "unknown" – which means they were far from us, and we were unaware of them. Hence the history of these societies only began when Europeans appeared – an epiphanal moment, qualitatively different from anything that had gone before, and culturally devastating. The historical difference with everything pre-colonial was power. Exposed and subjected to Western domination, the less powerful peoples were destined to lose their cultural coherence, as well as the pristine innocence for which Europeans – incomplete and sinful progeny of Adam – so desired them.

Sahlins (2005: 45)

Such issues are significant, as it is often too easy to portray cultural change and concerns over "authenticity" from a unidirectional perspective of tourism causing damage to communities or individuals without their own agency. Instead, processes of representation, identity, and heritage that are related to tourism are far more complex, as culture and heritage are inherently dynamic.

Culture changes, either gradually or rapidly, over time. Indeed, it is a system that changes with each new idea, new development, each new generation and each new interaction with other cultures and/or peoples. Past cultures lend themselves to conservation. Living cultures are based on legacies of the past, the ideas of the present and the hopes of the future. In trying to understand living cultures we must also understand their legacies from the past.

Kavaliku (2005: 23)

Indeed, as the next section discusses, the portrayal of the past in the Pacific is an area of significant contestation.

Portraying the Pacific through heritage and heritage tourism

The cultures of the Pacific have undergone significant rejuvenation since the 1970s. The reasons for this are manifold but particularly important is the connection of rediscovery of cultural identities as part of political activism and independence. As Wendt (1976: 60) stated:

this artistic renaissance is enriching our cultures further, reinforcing our identities, self-respect and pride, and taking us through a genuine decolonisation; it is also acting as a unifying force in our region. In their individual journeys into the Void, these artists, through their work, are explaining us to ourselves and creating a new Oceania.

Culture and heritage are obviously connected, but processes of commodification and cultural packaging are essential in turning culture into a heritage experience— whether for locals or for visitors. Therefore, heritage tourism is important for reinforcing the portrayal of emerging national identity and post-colonial image making in the Pacific through official sites of cultural knowledge such as museums and cultural centers (LeFevre 2007). However, the development of heritage "sites," particularly for tourism purposes, may itself change local understandings of heritage and representation. For example, Foana'ota (2007) noted that, in Solomon Islands traditional societies, keeping objects that modern museums tend to regard as important cultural materials was never practiced.

The idea of bringing together artefact collections representing different and diverse cultures and societies under one roof was a new phenomenon for the indigenous population of the islands. It came as part of a broad range of influences that were expected to affect the way people think about their own cultural heritage, and protect, preserve and promote them. The only time anything of cultural importance was seen in public was during special ceremonies or festivals. In some cases the items were destroyed or left to rot after the purposes for which they were made and used were no longer necessary or applicable.

Foana'ota (2007: 38)

Nevertheless, while traditional Pacific societies may have had quite different concepts of heritage than those of the colonial collectors, by the time the post-colonial period was under way, the island states of the Pacific were utilizing museums and cultural centers as mechanisms for place imaging and tourism development in a similar fashion to their European or North American counterparts (Cochrane 1999; Eoe 1990). As O'Hanlon (1994: 458) noted in the early 1990s, museums as "of interest in relation to the attempts by Pacific countries, internally diverse as they are, to consolidate themselves as nation-states and to make the variety of local experience speak to national ends." But such activities are fraught with difficulties between different perceptions and voices as to how the past should be interpreted through our present-day cultural, political, and economic needs. For example, Healey and Whitcomb (2006: 01.3) note:

> All museums in the South Pacific have had to engage with the past, par-
> ticularly the colonial past ... the pressures of history are most keenly felt
> as the clashes between the coloniser and colonised, indigenous and settler,
> reverberate continuously in the present.

A good example of issues of heritage interpretation and representation in the region is the case of the two main heritage institutions in New Caledonia that deal with the indigenous Kanak culture: the Museum of New Caledonia (MNC) and the Center Cultural Tjibaou (CCT). The institutions have a common focus but radically different interpretation.

The MNC, at one time referred to as the colonial museum, is New Cale-donia's oldest heritage institution. The MNC has a major collection of Kanak and Melanesian artifacts. However,

> the MNC is seen by Kanaks not only as a repository of tradition, the
> guardian of the indigenous ancestral culture, but also as a cemetery
> where objects are out of context. It is a significant place that holds powerful,
> special objects belonging to the dead, with most no longer used in society.
> That these objects are exposed to all visitors is disturbing to many
> Kanaks. Furthermore, the issue of Kanak objects being part of museum
> collections is still a sensitive one, tied as it is to colonisation and the loss
> of culture. Going to the museum is something most people consider with
> caution and some hesitancy.
>
> Tissandier (2006: 04.2–04.3)

In contrast, the CCT "has not inherited the same colonial history as the museum, yet it also has to engage with the various ethnic groups brought together by the colonial past and the political leaders' will to constitute a common heritage for all to share in the future" (Tissandier 2006: 04.3). The CCT is more of a living testimony to Kanak culture with a design drawn from that of traditional ceremonial huts, which enables visitors to directly experience various aspects of Kanak and Melanesian culture. Nevertheless, although presented in terms of architecture and interpretation as a break with the colonial past, some of the issues associated with the colonial legacy still remain.

> The Centre is funded by the French government. It stands in a country, New Caledonia, which while no longer fully a colony of France, is not yet an independent state. After a brutal colonial history and bloody struggle for independence, the Tjibaou Cultural Centre is both iconic and ambivalent in the Pacific.
>
> Losche (2007: 73)

In both the CCT and the MNC, as well as in other museums in the region, many issues connected with the violence associated with both colonialism and the development of new states in independent struggles clearly raise problems of interpretation and representation and are often not discussed at all or raised in oblique ways (Losche 2006). For example, Losche (2007) comments that many cultural centers in the Pacific "seem to conform to a narrative that ignores the fact that some, at least, have emerged from ruins and violent pasts and thus cultural centres seem rather clean spaces, uncontaminated by history." Nevertheless, she interestingly goes on to note that, "to some extent this makes sense in the Pacific region, where the development of positive identities, cultural pride and civil society takes precedence" (Losche 2007: 70). Therefore, while battlefield and war tourism may be an important component of heritage tourism to some parts of the region, especially in relation to Second World War sites (Panakera 2007), the extent to which the impacts of Japanese and American occupation is discussed in museum settings is extremely problematic, especially given the significance of visitor markets from these countries and the wish to promote positive perceptions of local hospitality (Figal 2008).

Nevertheless, while new cultural centres, particularly the CCT, are regarded positively as models for bringing culture and tourism together (Losche 2007), many have not been successful. For example, with the creation of custom-houses and cultural centers—"indigenous museums" in the Solomon Islands—one of the major causes of failure "was the belief that, by creating such centres, and putting objects inside, tourists would be attracted to visit them. Unfortunately, some of these centres were located in areas that were hard to get to and if visitors did come, high fees were often charged to enter and view the collections of artefacts they housed" (Foana'ota 2007: 41). Heritage tourism therefore continues to face some of the same locational and market issues that challenge all types of tourism (Burns 2005). More problematically, despite the wide-ranging enthusiasm for heritage tourism in the region, including more nature-based forms (Ringer 2004), the reality is that the size of the market that travels primarily for heritage experiences is likely to remain small because for many heritage is an adjunct to the more mainstream motivations of sun, sand, sea, and surf that are the focal point for the promotion of the region.

Conclusions

The creation of the Pacific in terms of otherness, identity and, clearly, as a destination lies in an understanding of the cultural politics of globalization. Cultural identity is "an ongoing process, politically contested and historically unfinished" (Clifford 1988: 9). Tourism in clearly inseparable from such cultural politics, which are:

> the struggles over the official symbolic representations of reality that shall prevail in a given social order at a given time. One could argue that they are the most important kind of politics, for they seek to control the terms in which all other politics, and all other aspects of life in that society, will take place.
>
> Ortner (1989: 200)

Nevertheless, tourism should not be observed in isolation as "tourism inevitably enters a dynamic context, and in the process contention over definitions of what is traditional and authentic become charged with a variety of additional meanings, as the range of interested parties increases" (Wood 1993: 63–64). As Harrison (2004) highlights, Christianity and colonialism are also crucial to understanding the changing cultural identities of the Pacific.

Culture, identity, and representations of heritage are continually being invented and reinvented by both insiders and outsiders (Hall 2009; Jolly 2005; Tunbridge and Ashworth 1996; Wood 2003). The significance of tourism "resides in the connections and disconnections it constitutes in the general processes of social change" (Hughes-Freeland 1993: 138). Tourism neither "destroys" culture nor does it ever simply "preserve" it. Instead, "In a fundamental sense, both today's and tomorrow's cultural tourists seek out not pre-development culture but the outcomes of different discourses and modes of development" (Wood 1993: 68). Therefore, as Tomlinson (1991: 28) observed, "we will have to problematise not just those cultural practices as 'modern', but the underlying cultural 'narrative' that sustains them: a narrative rooted in the culture of the (capitalist) West, in which the abstract notions of development or 'progress' are instituted as global cultural goals."

Despite the polyvocal communities of the Pacific, the state increasingly requires national heritage institutions to provide a marketable identity in which commerce, culture, and tourism converge (Tissandier 2006). Such an observation is extremely significant within the context of heritage and heritage tourism, which serves as the basis for much of the promotion and development of the Pacific as a tourist destination (Hall 1997; Hall and Page 1996; Harrison 2003, 2004). The representation and packaging of "how people live" for purposes of heritage tourism is something that should be seen as an ongoing interactive and contested process. The problem that heritage researchers often face is that "the political discourse of national culture and national identity *requires* that we imagine this process as 'frozen' and this is done via concepts like the 'national heritage' or our 'cultural traditions'. This 'freezing' conceals a complex historical process," in which sorting out the definitive features of a culture is highly problematic (Tomlinson 1991: 90).

> What we take to be "our culture" at any time will be a kind of "totalisation" of cultural memory up to that point. This totalisation will be a particular and selective one in which political and cultural institutions (the state, the media) have a globalization role ... as a consequence, "our culture" in the modern world is never purely "local produce", but always contains the traces of previous cultural borrowings or influence, which have been part of this "totalising" and have become, as it were, "naturalised".
>
> Tomlinson (1991: 91)

In many ways, the Pacific faces the same problems of heritage tourism experienced elsewhere in the developing world, with perhaps two differences. First, the countries of the region are extremely peripheral to the global economy in terms of geography, commerce, and politics, even more so than many other developing countries. Second, they are extremely vulnerable to the long-term effects of global environmental change. Heritage is therefore something under real material threat. Arguably the externally created exoticization of the Pacific, which has also been promoted more recently through local agency, is not only creating an unreal stereotypical image but, perhaps more seriously, is potentially masking the economic and environmental realities the region faces.

References

Bacchilega, C. (2007) *Legendary Hawai'i and the Politics of Place: Tradition, Translation and Tourism*. Philadelphia, PA: University of Pennsylvania Press.

Brown, G.H. (1988) *Visions of New Zealand: Artists in New Zealand*. Auckland: David Bateman.

Browne, C. (2006) *Pacific Island Economies*. Washington, DC: International Monetary Fund.

Browne, C. and Mineshima, A. (2007) *Remittances in the Pacific Region*. IMF Working Paper WP/07/35. Washington, DC: International Monetary Fund.

Burns, P. (2005) Ecotourism planning and policy "Vaka Pasifika"? *Tourism and Hospitality Planning & Development*, 2(3): 155–69.

Clifford, J. (1988) *The Predicament of Culture: Twentieth-Century Ethnography, Literature and Art*. Cambridge, MA: Harvard University Press.

Cochrane, S. (1999) Out of the doldrums: museums and cultural centres in Pacific island countries in the 1990s. In B. Craig, B. Kernot and C. Anderson (eds), *Art, Performance and Society*. Honolulu, HI: University of Hawai'i Press.

Connell, J. (1988) *Sovereignty and Survival: Island Microstates in the Third World.* Research Monograph No. 3. Sydney: Department of Geography, University of Sydney.

Connell, J. and Gibson, C. (2008) "No passport necessary": music, record covers and vicarious tourism in post-war Hawai'i. *The Journal of Pacific History,* 43(1): 51–75.

Department of Foreign Affairs and Trade (1994) *Trade Patterns – South Pacific.* Canberra: Department of Foreign Affairs and Trade.

Douglas, N. (1996) *They Came for Savages: 100 Years of Tourism in Melanesia.* Lismore: Southern Cross University Press.

Douglas, N. and Douglas, N. (1996) Tourism in the Pacific: historical factors. In C.M. Hall and S. Page (eds), *Tourism in the Pacific: Issues and Cases,* pp. 19–35. London: International Thomson Business Press.

Eoe, S.M. (1990) The role of museums in the Pacific: change or die. *Museum,* 41(1): 29–30.

eTurboNews (2008) Hawaii is running out of steam? Tourism slowdown felt across Hawaii. eTurboNews, January 28. Available from http://www.eturbonews.com/1043/tourism-slowdown-felt-across-hawaii (accessed April 1, 2008).

Everitt, T. (2008) Steady growth in South Pacific tourism. Regional News, April 4. Available from http://www.spto.org/spto/export/sites/spto/news/press/NewsLetter_Apr_2008_217.shtml (accessed May 1, 2008).

Farrell, B. (1982) *Hawaii: the Legend that Sells.* Honolulu, HI: University of Hawaii Press.

Figal, G. (2008) Between war and tropics: heritage tourism in postwar Okinawa. *The Public Historian,* 30(2): 83–107.

Foana'ota, L. (2007) The future of indigenous museums: the Solomon Islands case. In N. Stanley (ed.), *The Future of Indigenous Museums: Perspectives from the Southwest Pacific.* New York: Berghahn Books.

Fulford, T. and Lee, D. (2002) Mental travelers: Joseph Banks, Mungo Park, and the Romantic imagination. *Nineteenth-Century Contexts: An Interdisciplinary Journal* 24(2): 117–37.

Groves, R.H. (1995) *Green Imperialism: Colonial Expansion, Tropical Island Edens and the Origins of Environmentalism, 1600–1860.* Cambridge: Cambridge University Press.

Hall, C.M. (1997) *Tourism in the Pacific Rim: Development, Impacts and Markets,* 2nd edn. South Melbourne: Longman Australia.

—— (1998) Making the Pacific: globalization, modernity and myth. In G. Ringer (ed.), *Destinations: Cultural Landscapes of Tourism.* New York: Routledge.

—— (2009) Tourists and heritage: all things must pass. *Tourism Recreation Research,* 34(1).

Hall, C.M. and Page, S. (eds) (1996) *Tourism in the Pacific: Issues and Cases.* London: International Thomson Business Press.

Harrison, D. (ed.) (2003) *Pacific Island Tourism.* New York: Cognizant Press.

Harrison, D. (2004) Editor's introduction. Tourism in the Pacific Islands. *Journal of Pacific Studies,* 26(1 and 2): 1–28.

Healey, C. and Whitcomb, A. (2006) Experiments in culture: an introduction. In C. Healey and A. Whitcomb (eds), *South Pacific Museums: Experiments in Culture.* Clayton: Monash University Press

Hitchcock, M., King, V.T. and Parnwell, M.J.G. (1993) Tourism in South-East Asia: introduction. In M. Hitchcock, V.T. King and M.J.G. Parnwell (eds), *Tourism in South-East Asia.* London and New York: Routledge.

Honour, H. (1981) *Romanticism*. Harmondsworth: Penguin Books.

Hughes-Freeland, F. (1993) Packaging dreams: Javanese perceptions of tourism and performance. In M. Hitchcock, V.T. King and M.J.G. Parnwell (eds), *Tourism in South-East Asia*. London and New York: Routledge.

Jolly, M. (2005) Beyond the horizon? Nationalisms, feminisms, and globalization in the Pacific. *Ethnohistory*, 52(1): 137–66.

Kavaliku, L. (2005) Culture and sustainable development in the Pacific. In A. Hooper (ed.), *Culture and Sustainable Development in the Pacific*, pp. 22–31. Canberra: Asia Pacific Press at the Australian National University.

Laurier, E. (1993) "Tackintosh": Glasgow's supplementary gloss. In G. Kearns and C. Philo (eds), *Selling Places: The City as Cultural Capital, Past and Present*. Oxford: Pergamon Press.

LeFevre, T. (2007) Tourism and indigenous curation of culture in Lifou, New Caledonia. In N. Stanley (ed.), *The Future of Indigenous Museums: Perspectives from the Southwest Pacific*. New York: Berghahn Books.

Losche, D. (2006) Hiroshima mon amour: representation and violence in new museums of the Pacific. *South Pacific Museums* 1(1): 17.1–17.11.

—— (2007) Memory, violence and representation in the Tjibaou Cultural Centre, New Caledonia. In N. Stanley (ed.), *The Future of Indigenous Museums: Perspectives from the Southwest Pacific*. New York: Berghahn Books

Ministry of Tourism (1992) *General Information on Tourism in Fiji: Its Past and Future and Impact on the Economy and Society*. Suva: Ministry of Tourism.

National Tourism Office of Vanuatu (1990) *A History of Tourism in Vanuatu: A Platform for Future Success*. Port Vila: National Tourism Office of Vanuatu.

O'Hanlon, M. (1994) Reviewed work(s): Museums and Cultural Centres in the Pacific. Reviewed by S.M. Eoe and P. Swadling. *Man, New Series*, 29(2): 485.

Ortner, S.B. (1989) Cultural politics: religious activism and ideological transformation among 20th century sherpas. *Dialectical Anthropology*, 14: 197–211.

Pacific Islands Monthly (1990) Vanuatu's revival as untouched paradise. *Pacific Islands Monthly*, 60(3): 37–39

Panakera, C. (2007) World War II and tourism development in the Solomon Islands. In C. Ryan (ed.), *Battlefield Tourism: History, Place and Interpretation*. Oxford: Elsevier.

Ringer, G. (2004) Geographies of tourism and place in Micronesia: the "sleeping lady" awakes. *The Journal of Pacific Studies* 26(1–2): 131–50.

Rousseau, J. (1978) *The Social Contract and Discourses*, trans. G.D.H. Cole, rev. J.H. Brumfitt and J.C. Hall, Everyman's Library. London and New York: Dent & Dutton.

Sahlins, M. (2005) On the anthropology of modernity; or, some triumphs of culture over despondency theory. In A. Hooper (ed.), *Culture and Sustainable Development in the Pacific*, pp. 44–61. Canberra: Asia Pacific Press at the Australian National University.

Selwyn, T. (1993) Peter Pan in South-East Asia: views from the brochures. In M. Hitchcock, V.T. King and M.J.G. Parnwell (eds), *Tourism in South-East Asia*, London and New York: Routledge.

Smith, B. (1960) *European Vision and the South Pacific 1768–1850: A Study in the History of Art and Ideas*. Oxford: Clarendon Press.

south-pacific.travel (2008) Regional news. *Pacific Pulse: The Weekly Newsletter of south-pacific.travel*, July 11. Available from http://www.spto.org/spto/cms/news/press/

NewsLetter_Jul_2008_232.shtml#c97a553a-4ece-11dd-8607-a777dabd41f0 (accessed July 15, 2008).

State of Hawai'i (2005) *Hawai'i Tourism Strategic Plan 2005–2015*. Honolulu, HI: State of Hawai'i.

Sturma, M. (2002) *South Sea Maidens: Western Fantasy and Sexual Politics in the South Pacific*. Westport, CT: Greenwood Press.

Tissandier, M. (2006) Museums of New Caledonia: the old, the new and the balance of the two. In C. Healey and A. Whitcomb (eds), *South Pacific Museums: Experiments in Culture*. Clayton: Monash University Press.

Tomlinson, J. (1991) *Cultural Imperialism: A Critical Introduction*. Baltimore, MD: Johns Hopkins University Press.

Tunbridge, J.E. and Ashworth, G.J. (1996) *Dissonant Heritage: The Management of the Past as a Resource in Conflict*. Chichester: John Wiley & Sons.

UNCTAD (2007) *Handbook of Statistics 2007*. New York and Geneva: UN.

UNEP, UNWTO, and WMO (2008) *Climate Change Adaptation and Mitigation in the Tourism Sector: Frameworks, Tools and Practice* (M. Simpson, S. Gössling, D. Scott, C.M. Hall and E. Gladin). Paris: UNEP, University of Oxford, UNWTO, WMO.

Wendt, A. (1976) Toward a new Oceania. *Mana Review*, 1: 49–60.

Wood, H. (2003) Cultural studies for Oceania. *The Contemporary Pacific*, 15(2): 340–74.

Wood, R.E. (1993) Tourism, culture and the sociology of development. In M. Hitchcock, V.T. King and M.J.G. Parnwell (eds), *Tourism in South-East Asia*. London and New York: Routledge.

8 South Asian heritage tourism

Conflict, colonialism, and cooperation

Gyan P. Nyaupane and Megha Budruk

Introduction

This chapter focuses on heritage and tourism in South Asia, a region also known as the Indian subcontinent. Some scholars use the term "greater India" to describe the region. While this might be appropriate from an historical point of view, for political reasons, South Asia is the more accepted nomenclature among other South Asian countries (Mittal and Thursby 2006). Owing to the lack of consensus regarding the definition of South Asia, it is important to state which countries are being considered in this chapter. Several scholars and institutions define the region differently. For example, Myanmar and Iran are sometimes considered part of South Asia. This confusion exists because there is no clear geographical boundary between South Asia and other Asian regions, including Southeast Asia and the Middle East. Rather, only a geographical basis seems more appropriate by which to define the region from geopolitical, socio-political, and historical perspectives. For the purpose of this chapter, we refer to the South Asian Association for Regional Cooperation's (SAARC) definition. SAARC is an economic and political organization that provides a platform for the peoples of South Asia to work together in a spirit of friendship, trust, and understanding (SAARC 2008b). According to SAARC, South Asia encompasses Afghanistan, Bangladesh, Bhutan, India, Maldives, Nepal, Pakistan, and Sri Lanka.

Religion and politics have shaped the region in substantial ways. For example, two of the world's oldest religions, Hinduism and Buddhism, are rooted here. Although there is no exact date regarding when Hinduism, the world's oldest and third largest religion, was first practiced, its history can be traced back to approximately 5000 BCE when the Indus Valley/Harappa civilization flourished in the region (Singh 2006). Like Hinduism, Buddhism was founded more than 2,500 years ago and exported to other parts of the world, particularly Southeast and East Asia. Beyond religious influences, the region has been home to indigenous empires, as well as influences and threats from external forces (Najam 2003). Colonialism is one such force that has shaped South Asia. British, Portuguese, French, and Dutch colonial rule has left behind a cultural and historical legacy that is apparent even today in the architecture, food, celebrations, politics, educational, and judicial systems of these countries. As such, the region has a rich and varied cultural heritage that includes a large variety of tourist attractions such as temples, monasteries, monuments, forts, tombs, palaces, and a thriving and ever-changing, living culture. The region is also home to forty-nine of UNESCO's World Heritage Sites (WHS) (see Figure 8.1), most of which are located in India (27), followed by Sri Lanka (7), Pakistan (6), Nepal (4), Bangladesh (3), and Afghanistan (2). There are no WHS inscribed in Bhutan or the Maldives—the two smallest countries of the region. Of the forty-six sites, thirty-six are classified as cultural heritage and the rest are natural.

The diversity of cultures, climates, and topographies makes the region potentially important for economic development through tourism. Nonetheless, South Asia remains one of the poorest and most densely populated areas in the world. According to the *2006 Human Development Report* (UNDP 2006), South Asia is described as the world's second poorest region, following subSaharan Africa, in terms of per capita gross national product (GNP). Furthermore, almost half the world's poor (500 million) live in South Asia (UNDP 2006). The same report also declared South Asia the world's most illiterate realm, being home to some 50 percent of the globe's illiterate population. South Asia accounts for nearly a quarter of the earth's population (1.42 billion), but the region's combined gross national income (GNI) was only 2.14 percent of the global total (SAARC 2005). World Bank (2000) data on adult literacy, life expectancy, and population growth rates indicate that the South Asian states as a group lag behind the world in general, and other developing countries in particular. South Asia is also one of the fastest growing regions of the world, experiencing a population growth rate of some 1.7 percent annually, compared with the 1.2 percent growth rate the world over (SAARC 2005).

Religious and political legacies

Although the individual countries are very diverse, their socio-political ties make the region as a whole distinct from other areas of the world. Among the most important ties that characterize South Asia are religion and culture. The region has deeply rooted values, cultures, religions, art, and architectural styles, which need to be analyzed to understand heritage and tourism. Among the many religions historically practiced, Hinduism, a non-proselytizing faith, dominates. Approximately 80 percent of India's 1.12 billion inhabitants follow Hinduism. The religion is also prevalent in Nepal, with over 80 percent of that country's population adhering to Hindu teachings and practices. Although Buddhism was born and nurtured in South Asia and is the major religion in Sri Lanka and Bhutan, the spread of Islam during the Middle Ages created an Islamic population second in size to Hindus. Islam is the dominant religion in four of the region's eight countries: Pakistan, Bangladesh, the Maldives, and Afghanistan. Today, Pakistan has the world's second

India
1. Agra Fort○
2. Ajanta Caves○
3. Ellora Caves○
4. Taj Mahal○
5. Group of Monuments at Mahabalipuram○
6. Sun Temple, Konark○
7. Manas Wildlife Sanctuary▲
8. Kaziranga National Park⊕
9. Keoladeo National Park⊕
10. Churches and Convents of Goa○
11. Fatehpur Sikri○
12. Group of Monuments at Hampi○
13. Khajuraho Group of Monuments○
14. Elephanta Caves○
15. Great Living Chola Temples○
16. Group of Monuments at Pattadakal○
17. Sundarbans National Park⊕
18. Nanda Devi and Valley of Flowers National Parks⊕
19. Buddhist Monuments at Sanchi○
20. Qutb Minar and its Monuments○
21. Mountain Railways of India○
22. Mahabodhi Temple Complex at Bodhyaga○
23. Rock Shelters of Bhimbetka○
24. Champaner-Pavagadh Archaeological Park○
25. Chhatrapati Shivaji Terminus (formerly Victoria Terminus)○
26. Red Fort Complex○
27. Humayun's Tomb, Delhi○

Afghanistan
1. Minaret and Archaeological Remains of Jam●
2. Cultural Landscape and Archaeological
   Remains of Bamiyan Valley

Bangladesh
1. Historic Mosque City of Bagerhat○
2. Ruins of the Buddhist Vihara at Paharpur○
3. The Sunderbans⊕

Pakistan
1. Archaeological Ruins at Moenjodaro○
2. Buddhist Ruins of Takht-i-Bahi and Neighbouring
   City Remains at Sahri-i-Bahlol○
3. Taxila○
4. Fort and Shalamar, Gardens in Lahore●
5. Historical Monuments of Thatta○
6. Rohtas Fort○

Sri Lanka
1. Ancient City of Polonnaruwa○
2. Ancient City of Sigiriya○
3. Sacred City of Anuradhapura○
4. Old Town of Galle and its Fortifications○
5. Sacred City of Kandy○
6. Sinharaja Forest Reserve▲
7. Golden Temple of Dambulla○

Nepal
1. Kathmandu Valley○
2. Sagarmatha National Park⊕
3. Royal Chitwan National Park⊕
4. Lumbini, The Birth Place of the Lord Buddha○

Sites in Danger
Cultural Site●
Natural Site▲

Legend:
Type of Site
Cultural Site○
Natural Site⊕
Mixed Site▲

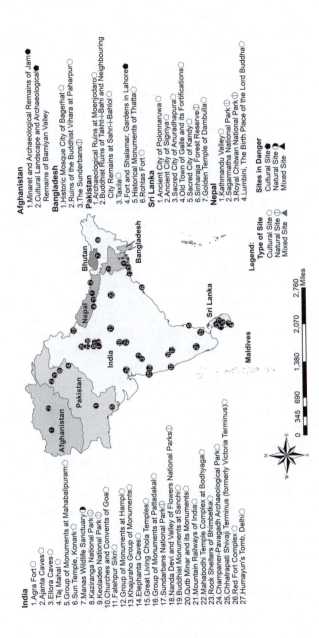

Figure 8.1 Developing countries and World Heritage Sites in South Asia

largest Muslim population. In addition, many other minority religions, such as Jainism and Sikhism, have contributed to the region's religious and cultural diversity. Christianity was introduced by European merchants and traders in the seventeenth and eighteenth centuries. During the colonial period and contemporary times, Christian missionaries were and are active, particularly in India. All these religious traditions have left a cultural footprint that forms much of the region's current heritage.

Religion also plays a central role in regional politics. Until 1947, much of South Asia—encompassing present-day India, Pakistan, and Bangladesh—was under British rule. When India gain independence, political agitation for a separate Muslim state led to the partition of British India into two nations: Hindustan, or land of the Hindus (India), and two wings of a Muslim majority Pakistan to the northwest and east of India. As a result, millions of Muslims moved to Pakistan while millions of Hindus and Sikhs moved to India leaving behind their places of religious and cultural importance. In 1971, East Pakistan separated from Pakistan and declared itself an independent Bangladesh. The religion-based separation of states as well as the dispute over Kashmir, a region claimed by both India and Pakistan, have been a major source of tension between the two countries. India and Pakistan have fought three wars over Kashmir since 1947. Similar connections between religion and politics can be seen in other nations. The Maldives was converted from a majority Buddhist nation to an Islamic state after the introduction of Islam in the twelfth century. The Maldives was then governed by Sultans under Dutch, Portuguese, and British rule until 1965, when it achieved independence. These religious and political contestations result in a set of heritages that are often contested and at the center of international debate.

Another major influence has been colonization. As already noted, the region has experienced British, Portuguese, and French rule, with significant implications regarding how heritage is preserved and portrayed. As stated above, prior to 1947, present-day India, Pakistan, and Bangladesh were colonized by Great Britain. Sri Lanka too was ruled by Britain until 1948. The Maldives attained independence from the British in 1965. Although Nepal and Bhutan were never formally colonized by Europeans, both countries lost significant portions of territory during the wars with British troops. The British East India Company, however, never returned the lands they had previously captured. As a result, residents of these areas, for example northeastern India, where a majority population of Nepalese live, are still fighting for their identity and a separate state.

Heritage and tourism in South Asian countries

A regional overview of tourism paints a dismal picture. Among the South Asian countries, Afghanistan's tourism industry has suffered the most because of civil wars and external invasions. Afghanistan served as a buffer state between the British and Russian empires until it won independence from the

British in 1919. Since then, Afghanistan has had a tumultuous political history with several coups and counter-coups. In 1973, a short-lived democracy was ended by a military coup, which was followed by a counter-coup by communist hopefuls in 1978. In 1979, the Soviet Union invaded the country and remained in power for ten years. With the fall of communism in 1989, Afghanistan regained political independence but, several years later, a civil war erupted and, in 1996, Afghanistan fell under the control of extremist Taliban leaders (Gohari 2000; The World Factbook 2008). More recently, following the September 11, 2001, terrorist attacks in the United States, American allied forces toppled the Taliban. Afghanistan is now undergoing major political, economic, and social transformation. Although Afghanistan has been able to draw tourists since the days of the Silk Road, the decades-long wars have all but halted tourism. With the downfall of the Taliban regime in 2001 the country reopened its borders to the world, and tourism has begun a slow process of recovery, with adventure tours from Europe being the primary new market (Lew *et al.* 2008).

Pakistan has been unsuccessful in tourism despite its natural and cultural resource potential, including Mount K2, the second highest peak in the world, the Indus Valley ruins, Arabian Sea beaches, and historic forts (Richter 1989, 1999). Geopolitical barriers and conflicts with its neighbors (China, Afghanistan, and especially India), as well as domestic political unrest (military coups and assassinations of political leaders) have led to a poor image. Additionally, travel to Pakistan from neighboring Western countries has been difficult owing to political tensions in the region. In the 1990s, most of Pakistan's international tourists were Indians who visited primarily to see family members and experience Hindu, Muslim, or Sikh holy places (Richter 1999). According to Richter, rather than cultivate this trend, Pakistan marketed itself as a Muslim tourism haven and generally focused on attracting the financially more lucrative Middle Eastern market. However, by 2004, the largest shares of international arrivals were from the United Kingdom (31 percent), United States (15 percent), and Afghanistan (10 percent). Even then, the most common reasons for visiting Pakistan were visiting family (56 percent), conducting business (21 percent), and holiday/recreation (15 percent) (Mehdi 2007). More recently, domestic and international political turmoil and an atmosphere of social conservatism have prevented tourism from becoming a major force in the country's economic development. In 2007, international tourist arrivals in Pakistan measured 351,000, representing a decrease of 6.6 percent from the previous year (Table 8.1).

Bangladesh, another Islamic nation, is home to the world's largest mangrove forest (Sunderbans) and the Royal Bengal Tiger. Its natural resources and beaches hold tremendous potential for eco-tourism and beachfront development. Unfortunately, natural calamities such as floods, disease epidemics, and cyclones adversely affect international arrivals. Recognizing the importance of tourism, Bangladesh established a National Tourism Policy in 1992 with the hope that tourism would alleviate poverty and improve socio-economic conditions

Table 8.1 Profile of South Asian countries

Country	Area (sq km)	Population	International tourists	Religious composition
Afghanistan	647,500	31,889,923 (July 2007 est.)		Muslim 99% Other 1%
Bangladesh	144,000	150,448,339 (July 2007 est.)	184,000*	Muslim 83% Hindu 16% Other 1%
Bhutan	46,500	634,982	10,000	Buddhist 75% Hindu 25%
India	3,287,590	1,129,866,154 (July 2007 est.)	2,400,000	Hindu 80.5% Muslim 13.4% Christian 2.3% Sikh 1.9% Other 1.9%
Maldives	300	369,031	675,889	Sunni Muslim
Nepal	147,181	28,901,790 (July 2007 est.)	418,000	Hindu 80.6% Buddhist 10.7% Muslim 4.2% Other 4.5%
Pakistan	803,940	164,741,924 (July 2007 est.)	351,000	Muslim 97% Other 3%
Sri Lanka	65,610	20,926,315	367,000	Buddhist 69.1%, Muslim 7.6% Hindu 7.1% Christian 6.2% Unspecified 10%

*2005 data.
Source: The World Factbook (2008).

in the country. As such, tourism was declared an industry and given priority in the subsequent annual five-year development plans. During the early 1990s, international tourist arrivals began steadily increasing and grew from 156,231 in 1995 to 184,000 in 2005. Tourism planning has remained largely government controlled, resulting in the planned development of tourism products partly dependent on the socio-economic conditions of the country.

During the middle years of the twentieth century, Sri Lanka was not a popular tourist attraction for the Western world. Economic reforms and tourism strategies by the government in the 1970s made Sri Lanka attractive to international aid agencies and tourists. During the years following these reforms, Sri Lanka became the world's leading per capita aid recipient (Arunatilake *et al.* 2001) and experienced a significant growth in its tourism industry (Bandara 1997). By 1982, tourism increased at an average rate of 22 percent annually and tourist arrivals exceeded 400,000. At the same time, tourism-related foreign exchange earnings and employment increased by more than 200 percent and 100 percent, respectively (Bandara 1997), indicating the importance of tourism to Sri Lanka's economy. Subsequent years saw ongoing tensions between the majority Sinhalese and minority Tamil groups erupting into armed conflict. This led to adverse consequences in every sphere of social and economic life. Tourist numbers dwindled to a low of 183,000 in 1988, and foreign exchange earnings and employment levels dropped significantly. Recovery of tourism during the 1990s was slow and, by the end of that decade, the industry had weakened enough to have considerable negative repercussions on the economy. Sri Lankan tourism development has taken place in a vacuum, largely ignoring the political environment that was necessary to sustain it. Rather than contributing toward political stability in the country, tourism was the first casualty of the war (Richter 1999). In December 2004, the large tsunami that tragically affected much of Southeast and South Asia devastated much of the tourism infrastructure of Sri Lanka, and many tourists and Sri Lankans lost their lives. Since that time, tourism has recovered well and a new image is re-emerging (de Sausmarez 2005; Jayasuriya *et al.* 2005).

Within the region, tourism looks promising for the Maldives. The Maldives was not a popular destination among international tourists until 1972, when the first resorts were established (Shareef and McAleer 2007). Although the country is the region's smallest, it is the most economically dependent on tourism. In 2007, a total of 675,889 tourists visited the Maldives, a figure 8.5 percent higher than the previous year (Ministry of Tourism and Civil Aviation Maldives 2008). Europe is the most important source of tourists, with Europeans accounting for more than three-quarters of total international arrivals (Shareef and McAleer 2007). Today, tourism is the largest economic sector in the Maldives, comprising some 30 percent of the country's GDP and 40 percent of total tax revenue (Ministry of Tourism and Civil Aviation Maldives 2008).

India, the region's largest state and second most populous country in the world, attracts about half of South Asia's tourists. Nonetheless, despite its vast and diverse history, geography, and cultural landscape, India's share of international tourism was only 0.4 percent of the world's total between 1957 and 1997 (Jithendran and Baum 2000). Government involvement in tourism can be traced to 1958 when the Ministry of Tourism was created. This entity was renamed the Ministry of Tourism and Culture in 2000 and continues to be the primary agency for tourism development and promotion in India. Despite its small share in the international tourist market, India is among the leading tourist destinations in South Asia. Most tourists arrive from the United Kingdom and the United States, many of whom are a part of the Indian diaspora returning to the homeland to visit friends and relatives. A liberal and growing economy in the 1990s has boosted tourism. Foreign exchange receipts from tourism increased from US$1.5 billion to about US$3.0 billion by the end of the decade. Subsequently, India promoted itself as a tourist destination by declaring 2000 as the Explore India Millennium and launching the "Incredible India" campaign in 2002. Both these strategies were designed to attract the growing international tourist sector. According to the Center for Monitoring the Indian Economy (CMIE), tourism has ranked fourth in foreign exchange revenue since 1999 (Bhattacharya and Narayan 2005). However, like other countries in the region, India's tourism industry is impacted by sporadic religious and political violence, lack of safety and infrastructure, and a poor image abroad.

Nepal and Bhutan are the only two countries in the region not to have been colonized by European metropoles. Nepal was ruled by an autocratic regime for over a century until 1951. Since the demise of the Rana regime and the opening of the country to the rest of the world, Nepal has experienced a gradual growth in tourist arrivals from 6,000 in 1962 to almost 500,000 arrivals in 2000 (Ministry of Culture Tourism and Civil Aviation 2006). Tourism has been considered among one of the most important sectors, as it accounts for 4 percent of the country's GDP and creates more than 250,000 direct and indirect jobs (Nepal Tourism Board 2001). Most tourists visiting Nepal are from Asia (51.9 percent), followed by Europe (32.6 percent) and North America (9.3 percent). The steady growth of Nepal's tourism industry in the 1980s and 1990s was interrupted by a Maoist insurgency in the early 2000s. Currently, Nepal is undergoing major political transformations with the king being dethroned in 2006 and the country being declared a secular state rather than a Hindu kingdom. Nepal had a constituent assembly election in April 2008 to elect a body to write a new constitution for the country. Nepal's tourism is going through a fast recovery, thanks to recent political developments that have brought Maoist rebels into mainstream politics.

Unlike other South Asian countries, Bhutan has a low population density (25.8 persons/sq km). Bhutan's tourism industry is unique as the country is promoting high-yield, low-impact tourism. Although the country is visited by small tourist numbers (just over 10,000 international and 17,000 Indian/ regional tourists), tourism has played a significant role in the economy. It is the largest foreign exchange earner and second largest revenue generator. It also provides some 9,000 direct and indirect jobs and is the largest employer in the private sector (National Portal of Bhutan 2008). Tourism has been largely controlled by the government through relatively high daily tariffs and pre-established itineraries (Zurick 2007). Travelers can only undertake package tours, for which the government charges a US$200 minimum daily tariff for each person in a group of four or more. Approximately a third of the daily tariff (US$65) goes to the government, and the remaining amount covers services such as food, accommodations, and transportation. The future of tourism and heritage in Bhutan will be influenced by the new democratic government that was elected to power in March 2008. Additionally, Bhutan is the only country in the world to measure and include happiness in an attempt to define quality of life. The gross national happiness (GNH) index is based on the premise that the overall happiness of the Bhutanese, rather than commonly used development measures such as GDP or human development index, is a more holistic measure of human development. The concept of GNH was first described by Bhutan's king in 1972 and is based on non-material goods such as religious traditions, values, cultures, spiritual fulfillment, and quality of the environment (Zurick 2007). Thus, the concept of GNH underscores the importance of preserving heritage and culture.

Despite low numbers of tourist arrivals to the region in the last century, South Asia is experiencing one of the highest average annual growth rates of arrivals in recent years (6.4 percent compared with the world average of 3.6 percent since 2000) (UNWTO 2007). The region recorded 11 percent growth in 2006 alone, compared with 2005. These numbers do not include most of the regional arrivals, as many South Asian countries have open border policies among themselves making passports and visas unnecessary when visiting bordering countries. For example, Indians do not require travel documents to visit Nepal, Bhutan, or Bangladesh, and vice versa. Although Bhutan has a mandatory daily tariff for foreign tourists, Indians, Bangladeshis, and Maldivians do not require visas and are not required to pay the tariff to visit Bhutan. Except Bhutan, where records of regional tourists are kept, tracking overland intraregional tourist numbers is nearly impossible in other South Asian countries.

Domestic and intraregional tourism in South Asia is still predominantly for pilgrimage rather than for leisure, especially for older people, although the commercialization and secularization of pilgrimage trips are rising (Edensor 1998; Sharpley and Sundaram 2005; Singh 2004). Hindu pilgrimage is translated as *tirtha yatra* (tour of sacred fords or crossings) in Sanskrit, an ancient language of the Indian subcontinent. In Sanskrit, *tir* means edges of the water body and *yatra* means a journey. Traveling to bathe in holy rivers, pray at sacred places, perform rituals, attend major religious events for the purpose of redemption and attaining religious merit have been common practices since early civilization, dating back to the Indus valley civilization (Singh 2004). Hindu pilgrimage has been institutionalized for over a thousand years by creating a circuit of four major holy sites: Badrinath in the north, Dwarik in the west, Jagannath Puri in the east, and Rameshwaram in the south. The function of pilgrimage goes beyond the personal religious salvation to family responsibility and social function. Traditionally, sons who took their aged parents on pilgrimage were considered obedient and model sons. Hindu pilgrimages were traditionally undertaken on foot, but with the spread of motorized vehicles, the method of travel and its impacts on the sites have been enormous. Many tourist sites in South Asia are symbolic sacred locations and centers of religious practice both locally and nationally.

Other forms of heritage tourism attractions that are even more varied and unique to the region are living cultures. The region is not just rich in physical assets, such as archeology, historic buildings, monuments, and temples, but people still practice ancient traditions and religious rituals, which draw many regional and Western tourists to South Asia (Singh 2004). While religious events are declining and heritage sites are becoming lifeless in many parts of the world, participation in these events is increasing in South Asia, and more and more tourists are attracted to experience these events and participate in people's everyday rituals. For example, Kumba Mela, a Hindu religious event that occurs four times every twelve years, is attended by 70 million people. This event is the largest gathering of its kind in the world and reflects thousands of years of history and tradition in India (Singh 2006).

Heritage and tourism issues

Even though tourism in the region is growing, South Asia receives only a small share of the world's international tourist market, represented by just 1 percent of the 846 million international trips and 1.6 percent of the US$733 billion world tourism receipts in 2006 (UNWTO 2007). This small share is due in part to negative stereotypes, which have perpetuated images of mass poverty, squalor, and disease. As noted above, other deterrents in the region are political instability and conflict, social disruption, and civil war. The depressing picture of tourism reflects that, although the region has a rich heritage, which needs preservation, governments have other priorities such as poverty alleviation, education, and health. For people struggling to survive, heritage conservation is not a high priority. As such, issues related to heritage and tourism in South Asia are intertwined with economy, history, geography, demographics, politics, and religion. Understanding these issues needs a holistic approach.

South Asia has been characterized as one of the most fragile regions in the world because of domestic political unrest and conflicts with neighboring countries. This is perhaps the largest issue adversely affecting heritage and tourism. The World Bank has reported that South Asia has the world's largest conflict-affected population (over 71 million in Afghanistan, Sri Lanka, Nepal, and Bhutan) (World Bank 2006). Most conflicts in South Asia are related to ethnicity and ethnonationalism (Tambiah 1996). In some instances, conflicts have resulted in the destruction of the heritage of one ethnic or religious group by another. Often, heritage sites, which are surrounded by a community of "others," are in particular danger (Nyaupane 2009). As discussed in the previous section, when people migrate because of either conflict or politics, they are forced to leave behind temples, buildings, and other forms of their tangible heritage. As a consequence, many heritage sites are surrounded by those for whom these sites have little meaning. Several such Buddhist sites exist in Afghanistan, Pakistan, and Bangladesh, which are predominantly Muslim countries and where negligible populations of Buddhists live (Nyaupane 2009). Perhaps the best known example of such a site is Bamyan Valley, Afghanistan, valued for the world's tallest Buddha statues (55 and 38 meters tall). Bamyan lies on the Silk Road, a historic caravan route linking China with Europe. The valley was believed to have been an active Buddhist religious center from the second century to the ninth century, when the Islamic invasion occurred. These statues were destroyed by the Taliban in 2001, for several reasons, one of which was to retaliate against the international community that had invaded and placed sanctions on the country (Ashworth and van der Aa 2002). Although the cultural landscape and archeological remains of the valley are protected and the site was subsequently inscribed on UNESCO's World Heritage List in 2003, it is still listed as an endangered site with the hope that something might yet be done. Similar attacks have threatened historic places in Pakistan. Suspected pro-Taliban militants attacked an ancient rock carving of a seated Buddha located near Janabad in northwest Pakistan in September 2007 (UNESCO 2008). The attacks partially damaged the seven-meter-high, seventh-century statue. During the Sri Lankan civil war between minority Hindu Tamils and majority Buddhist Sinhalese, Sri Dalada Maligawa in Kandy (the Temple of Tooth), a UNESCO World Heritage Site, was bombed in 2001 by the Liberation Tigers of Tamil Eelam (LITTE). Fortunately, only the entrance to the temple was harmed and no damage was done to the core area where a tooth of Buddha is believed to be kept (Coningham and Lewer 1999). Similarly, many Hindu temples were burned down and closed during the civil war in Sri Lanka in the 1990s and early 2000s. Conflicts such as these are detrimental not only to the heritage but to tourism as well. As of June 24, 2008, four of seven South Asian countries, including Afghanistan, Pakistan, Nepal, and Sri Lanka, were on the US State Department's travel warning list (US Department of State 2008).

A second issue is that of the internationalization of heritage through external forces such as globalization. Although globalization is explained mostly in economic terms such as internationalization of trade, labor, and production, it has direct implications for heritage conservation. International assistance in the protection of heritage is sometimes required, as in the case of Bamyan Valley. However, this may at times be perceived as problematic if local communities are excluded from the decision-making process involving heritage sites that remain meaningful to them. What may sometimes result is the internationalization of heritage sites, referring to the conversion of local and national heritage into global heritage. In such cases, external economic, political, and ideological interests have the potential to supersede local interests. At times, Western meanings are brought to heritage through a history of colonization. Other times, international organizations such as UNESCO and the International Council on Monuments and Sites (ICOMOS) (which are involved in heritage conservation) represent Western values and follow international principles of conservation that are not sensitive to local values and issues. The Taj Mahal, India's most popular tourist icon, is a vivid example of both of these. Despite the presence of several older and more impressive monuments, the Taj was deemed important to preserve in the 1860s by European archeological survey members, thus establishing its worth as a legitimate part of Indian heritage (Cohn 1984; Pal 1989). For Westerners, the Taj is a beautiful monument made of marble and based on a romantic story from the seventeenth century. As such, other and perhaps more significant monuments have not received the attention they deserve. Sometimes, tourism products ignore the unique characteristics of the destination and replace them with images as conceived by Western tourists (Britton 1979). More recently, the central government of India and UNESCO considered locals as a disruption and excluded residents from the planning and decision-making process regarding the Taj Mahal.

Another issue relates to heritage "belonging" to several groups, and the resulting contested meanings and conflict. Again, the Taj Mahal is illustrative of how the monument's secular and religious symbolism is blurred by different groups. Among these meanings are the colonial, sacred, and national representations of space (Edensor 1998). Colonial or Western meanings of the Taj Mahal stem from the need for Western tourists to experience the "other" or that which is outside their regular environment. Contemporary guides and tour books follow this fascination by including the Taj as a "must see" in package tours to India. A second manifestation of meaning is that of the sacred. For some Muslims, the Taj Mahal, a Muslim mausoleum, holds sacred meaning with thousands visiting the monument and considering it a pilgrimage site. A mosque located within the complex is still visited by Muslim worshippers. The Taj is open only for Muslim worshippers on Fridays and closed for others. For Hindus, Buddhists, and Jain pilgrims en route to sacred centers close to the Taj, the monument serves as a draw owing to its historical and cultural significance. In fact, over half of the domestic visitors to the Taj represent this latter group. A third meaning ascribed to the Taj is that of a national symbol. Partly as a result of colonial representations of Indian heritage, the Taj Mahal is now considered a monument that epitomizes Indian national identity. Different meanings result in different "owners" of heritage. While in the case of the Taj, conflict resulting from different meanings has remained subtle, in other cases, ethnic and religious conflict over heritage may spill over into riots, which involve violence, homicide, and destruction of heritage and property. For instance, the dispute over the site of the Babri Mosque as being the original site of the Ayodhya Temple in India resulted in loss of both Muslim and Hindu lives and the destruction of the sixteenth-century Babri Masjid in 1992. Other examples include increasing radical Hindu nationalism and anti-Muslim violence that resulted in the Gujarat massacre in India in 2002. In Bangladesh, violence against the country's Hindu minority has resulted in thousands of Hindus seeking refuge in India (Cady and Simon 2007). Similarly, in Bhutan, about 110,000 Hindus of Nepali origin were expelled in the name of cultural preservation and national unity, and became refugees in Nepal. Many of these are in the process of third country resettlement in the USA, Australia, Canada, Norway, the Netherlands, New Zealand, and Denmark.

Similarly, there are some cases in South Asia where important heritage sites are ignored by locals. Although the conflicts are very subtle compared with the Bamyan Valley, Afghanistan, governments and international agencies need to pay attention to these sites. Lumbini, Nepal, the birthplace of Lord Buddha, is one of them. Lumbini has been lost and discovered many times over the centuries. There has been much speculation as to why Lumbini remained abandoned and forgotten for centuries. Possible causes include the influence of other religions, Muslim incursions and domination, the restoration of Hinduism, and natural disasters including earthquakes, floods, and malaria (Bidari 2004; Falk 1998; Kate and Ulrike 2005; Pandey 1985). Like Lumbini, many sites in northern India, including Sarnath, where Buddha delivered his first sermon, and Kushinagar, where Buddha died, were also abandoned and ignored (Nyaupane 2009). Currently, these sites are surrounded by Hindus and Muslims, which can be a major challenge for heritage preservation as the locals do not see this heritage as part of their own spiritual past.

While most of the above discussion centers around a dissonance of values and meanings associated with heritage, the major issue faced by several South Asian countries is the lack of involvement of local communities in heritage conservation, even though this heritage remains significant to the said community for spiritual, historical, or economic reasons. Often, economic pressures faced by such communities lead to negative impacts on these sites. For example, heritage sites and their surroundings are often encroached upon by locals for agriculture and housing. A lack of funding for heritage conservation often translates into poor enforcement of rules and regulations, subsequently compounding the matter. Heritage tourism has the potential to alleviate this issue through economic development, as well as the creation of a sense of pride among the community concerning its heritage.

Opportunities

The increasing importance of culture and heritage as part of the tourist experience has been documented, with the World Tourism Organization (UNWTO) estimating that cultural tourism accounts for nearly 40 percent of all tourist trips (McKercher 2002). Additionally, the cultural heritage tourism market is one of the fastest growing sectors within the tourism industry (Aluzua *et al.* 1998). Thus, as a region desperately seeking solutions to poverty reduction, South Asia is uniquely poised to tap into the growth and volume of the cultural and heritage tourism market.

The UNWTO highlighted several reasons that make tourism an especially viable option for economic development in the developing countries, which can be true for South Asia (UNWTO 2007). Among these is the fact that tourism is consumed at the point of production, which means tourists have to travel to the communities to consume the product (experience). This provides an opportunity for countries to gain economically by tourists injecting new money into the economy. The economic importance of tourism to the Maldives is an especially notable example, with about a third of its GDP dependent on tourism. Tourism is also an important source of foreign exchange for several countries such as Nepal, Bhutan, and India. The second reason highlighted by the UNWTO is that visits to heritage and cultural settings generate employment and income for communities and contribute to the conservation of these sites. Third, because of its diverse nature, tourism provides flexible and part-time job opportunities that complement other sources of livelihood. Fourth, tourism is labor intensive and is supported by a skilled and unskilled labor force, making it a unique source of employment in countries that have high populations and low levels of literacy. South Asian countries exhibit both these population characteristics. For example, India has the world's second largest population but also a 65 percent literacy rate (75 percent of men and 54 percent of women) (Census of India 2008). Unskilled labor can provide a unique opportunity for these populations (especially women) to gain economic freedom. Next, tourism inculcates a sense of ownership and pride for natural and cultural resources in the communities depending on these resources. As seen earlier, communities that are not culturally or historically invested in local heritage are less likely to care for and protect artifacts from the past. Finally, the infrastructure required by the tourism industry, such as roads, airports, water supply, and sanitation, benefits all sectors of society.

Beyond these economic opportunities, tourism, especially cultural heritage tourism, can potentially increase cross-border and cultural understanding among tourists and destination communities. In other words, heritage tourism can provide a unique basis for countries such as India and Pakistan, which share deeply entwined histories, to build common ground, patch relations, and reduce conflict. It is apparent that visitors from either country who cross borders primarily do so to visit family and/or friends. Recognizing this, in February 2006, India and Pakistan launched a weekly train service between the two countries allowing the first people-to-people contact since both countries gained independence in 1947. While such exchanges may increase cross-border understanding, reduce conflict, and simultaneously encourage regional stability, cooperation on a larger scale is vital for the economic progress of the region as a whole. Although the need for regional cooperation to promote the exchange of ideas and the development of new economic and cultural links was felt for a long time, it did not materialize until 1985, when the South Asian Association for Regional Cooperation (SAARC) was established. Initially, SAARC included seven member countries: Bangladesh, Bhutan, India, the Maldives, Nepal, Pakistan, and Sri Lanka. After the abolition of the Taliban in 2001, Afghanistan joined the alliance. Since its inception, SAARC has recognized tourism as one of the fifteen key areas of cooperation, realizing the importance of economic development through tourism. Although tourism is largely seen as a private sector industry, its development, promotion, and ultimate success demands cooperation between governments. The second SAARC summit held in Bangalore, India, in 1986 emphasized the need for promoting people-to-people contact between member countries through tourism (SAARC 2008a). In 1990, the fifth SAARC summit, held in Male, Maldives, endorsed the proposal for institutionalized cooperation among the tourist industries in the region to attract more tourists from outside the region (Timothy 2003). Tourism has also been an important area of discussion for most of the subsequent summits. At the twelfth summit in Islamabad in January 2004, the leaders commemorated the twentieth year of the establishment of SAARC, declaring the year 2005 "South Asia Tourism Year." All member states individually and jointly organized many special events to celebrate it (SAARC 2008a). In addition, the Tourism Working Group within SAARC is implementing several programs to promote tourism in the region including the printing of a SAARC Travel Guide, the production of a documentary movie on tourism in SAARC, and promoting the sustainable development of eco-tourism, cultural tourism, and nature tourism.

There is also a growing diaspora tourism in South Asia as increasing numbers of South Asians living in the Western world visit their homelands to reconnect with their roots, history, culture, and heritage. Although diaspora exists in each of the South Asian countries, the most documented one is that of India (Hannam 2004). It is estimated that an average of 17 million people from India live worldwide. These people carry varied experiences, religions, languages, ethnicities, and socio-economic statuses giving rise to a splintered and heterogeneous group that is yet held together by a strong sense of identity in their adopted homelands. Clarke *et al.* (1990) identified five different types of Indian migrants. These include emigrants from British India, commercial migrants, migrants to the Middle East, migrants to the US, UK, and Canada during and after the 1980s, and migrants to other South Asian countries such as Pakistan and Bangladesh. The informal ties between these groups and their heritage in India provide a tremendous growth opportunity for the South Asian tourism industry. A report on the Indian diaspora published by the Ministry of External Affairs, India, in 2001 acknowledges the economic potential of this group, primarily contributing to tourism through capital investment in projects such as resorts and other visitor attractions. Unfortunately, the complexity of the diaspora and geopolitical concerns have resulted in the lack of a unified strategy to engage in diaspora tourism (Hannam 2004). However, recent changes to this ambivalence are apparent. The Ministry of External Affairs report mentioned above suggested that the Indian government encourages members of the Indian diaspora to visit the country for tourism purposes. Additionally, the report recommended the establishment of an annual "Family of India Day" on January 9. This date was selected because of its symbolic national significance; on this day in 1915, Mohandas Karamchand Gandhi (known as Mahatma Gandhi) returned to India from South Africa to participate in India's freedom struggle. The *Pravasi Bharatiya Divas*, as the day is known, has been celebrated from the year 2003 onwards and is sponsored by the Ministry of Overseas Indian Affairs and an association of Indian business organizations. Despite shortcomings associated with these initial attempts to re-establish contact with the diaspora, it nevertheless provides an opportunity for the Indian diaspora to be recognized as having a legitimate connection with India. India is not the only country with a diaspora. Certainly, Pakistan, Nepal, Bangladesh, Sri Lanka, and other South Asian countries have diasporas that are waiting to be contacted as well.

Conclusions

It is clear that heritage tourism remains an important economic development tool for South Asia. The region is home to numerous world-renowned tourist attractions that remain important not only to the country in which they are situated but to the world at large. The diverse architecture and other manifestations of heritage, including living culture, in South Asia serve as a draw for tourists seeking authentic cultural and heritage experiences. The region's major challenge is poverty, and heritage tourism can potentially eradicate or reduce poverty through economic stimuli. To link heritage with the economic needs of the people, carefully planned and managed heritage tourism can help provide jobs and other economic opportunities to residents. The economic values can drive these communities toward preservation of sites and living culture and an increase in community pride regarding heritage. International organizations and governments should be sensitive to the locals' values and needs while developing and preserving heritage.

Some of the legacies of the past have been destroyed, mostly because of anthropocentric factors such as wars and conflicts, and fires, and by natural disasters such as tsunamis, floods, and earthquakes. What remains should be preserved through the cooperation of local stakeholders, governments, and international agencies. Many heritage sites in South Asia are threatened because of a lack of pride or ownership in them by the community living adjacent to these sites (Chakravarti 2008; Nyaupane 2009). While the role of government and international agencies is important, linking residents to these places is equally important. It appears that the governments are more interested in promoting various forms of leisure tourism over pilgrimage tourism in South Asia, as the economic gains from pilgrimage seem to be less than those from other forms of tourism. Governments and the tourism industries should recognize the potential of religious tourism as a form of heritage and pay more attention to its economic and social role.

South Asia is inherently very rich in cultural and ethnic diversity, which has unfortunately been a source of religious and ethnic conflict. This cultural and ethnic diversity can potentially be turned into an asset that characterizes the region and plays an important role in its economic development. Promotion of heritage-based regional tourism can also be a tool for peace building through people-to-people contact. There are stereotypes based on religion, ethnicity, race, language, and culture that can be alleviated or reduced by encouraging regional or domestic tourism.

Although South Asia has varied cultural, heritage, and natural attractions to draw intraregional and international tourists, the region needs to upgrade its tourism amenities such as accommodation and transportation. There is a lack of basic service quality in the tourism and hospitality sector, which needs to be improved by providing more tourism and hospitality education and training. Although some countries and destinations in South Asia are safer than others, overall there is a strong perception among Western tourists that the region is unsafe to visit. The governments and tourism industries of the region must work together to plan and promote sustainable cultural and heritage tourism for the good of all the countries in South Asia.

References

Aluzua, A., O'Leary, J.T. and Morrison, A.M. (1998) Cultural and heritage tourism: identifying niches for international travelers. *The Journal of Tourism Studies*, 9(2): 2–13.

Arunatilake, N., Jayasuriya, S. and Kelegama, S. (2001) The economic cost of the war in Sri Lanka. *World Development*, 29(9): 1483–500.

Ashworth, G.J. and van der Aa, B.J.M. (2002) Bamyan: whose heritage was it and what should we do about it? *Current Issues in Tourism*, 5(5): 447–57.

Bandara, J.S. (1997) The impact of the civil war on tourism and the regional economy. *South Asia*, 20: 269–79.

Bhattacharya, M. and Narayan, P.K. (2005) Testing for the random walk hypothesis in the case of visitor arrivals: evidence from Indian tourism. *Applied Economics*, 37: 1485–90.

Bidari, B. (2004) *Lumbini: A haven of sacred refuge*. Kathmandu: Hill Side Press (P) Ltd.

Britton, R.A. (1979) The image of the third world in tourism marketing. *Annals of Tourism Research*, 6(3): 318–29.

Cady, L.E. and Simon, S.W. (2007) Introduction: reflections on the nexus of religion and violence. In L.E. Cady and S.W. Simon (eds), *Religion and Conflict in South and Southeast Asia*, pp. 3–20. New York: Routledge.

Census of India (2008) Available from http://www.censusindia.gov.in/ (accessed April 20, 2008).

Chakravarti, I. (2008) Heritage tourism and community participation: a case study of the Sindhudurg Fort, India. In B. Prideaux, D.J. Timothy and K.S. Chon (eds), *Cultural and Heritage Tourism in Asia and the Pacific*, pp. 189–203. London: Routledge.

Clarke, C., Peach, C., and Vertovec, S. (eds) (1990) *South Asians Overseas: Migration and Ethnicity*. Cambridge: Cambridge University Press.

Cohn, B. (1984) The census, social structure and objectification in South Asia. In *An Anthropologist among the Historians and Other Essays*, pp. 224–56. Delhi: Oxford University Press.

Coningham, R. and Lewer, N. (1999) Paradise lost: the bombing of the Temple of the Tooth – a UNESCO World Heritage Site in Sri Lanka. *Antiquity*, 73: 857–66.

de Sausmarez, N. (2005) The Indian Ocean tsunami. *Tourism and Hospitality Planning and Development*, 2(1): 55–59.

Edensor, T. (1998) *Tourists at the Taj: Performance and Meaning at a Symbolic Site*. London: Routledge.

Falk, H. (1998) *The Discovery of Lumbini*. Occasional paper 1. Kathmandu: Lumbini International Research Institute.

Gohari, M.J. (2000) *The Taliban Ascent to Power*. Oxford: Oxford University Press.

Hannam, K. (2004) India and the ambivalences of diaspora tourism. In T. Coles and D. J. Timothy (eds), *Tourism, Diasporas and Space*. New York: Routledge.

Jayasuriya, S., Steele, P. and Weerakoon, D. (2005) *Post-Tsunami Recovery: Issues and Challenges in Sri Lanka*. Tokyo: Asian Development Bank Institute.

Jithendran, K.J. and Baum, T. (2000) Human resources development and sustainability – the case of Indian tourism. *International Journal of Tourism Research*, 2(6): 403–21.

Kate, M. and Ulrike, M.-B. (2005) The local impact of under-realisation of the Lumbini master plan: a field report. Available from http://goliath.ecnext.com/coms2/gi_0199–5686460/The-local-impact-of-under.html (accessed October 3, 2007).

Lew, A.A., Hall, C.M. and Timothy, D.J. (2008) *World Geography of Travel and Tourism: A Regional Approach*. Oxford: Butterworth Heinemann.

McKercher, B. (2002) Towards a classification of cultural tourists. *International Journal of Tourism Research*, 4(1): 29–39.

Mehdi, Z. (2007) Exploiting tourism potential. *Management Account*, 16(4): 17–30.

Ministry of Culture Tourism and Civil Aviation (2006) *Nepal Tourism Statistics 2006*. Kathmandu: Government of Nepal, Ministry of Culture, Tourism and Civil Aviation.

Ministry of Tourism and Civil Aviation Maldives (2008) Tourism in Maldives. Available from http://www.tourism.gov.mv (accessed April 12, 2008).

Mittal, S. and Thursby, G. (2006) Introduction. In S. Mittal and G. Thursby (eds), *Religions of South Asia: An Introduction*, pp. 1–11. London: Routledge.

Najam, A. (2003) *The Human Dimensions of Environmental Insecurity: Some Insights from South Asia*. Washington, DC: The Woodrow Wilson Center.

National Portal of Bhutan. (2008) Fact file. Available from http://www.bhutan.gov.bt/government/aboutbhutan.php (accessed April 4, 2008).

Nepal Tourism Board (2001) *Annual Report 2001*. Kathmandu: Nepal Tourism Board.

Nyaupane, G.P. (2009) Heritage complexity and tourism: the case of Lumbini, Nepal. *Journal of Heritage Tourism*, 4.

Pal, P. (1989) *Romance of the Taj Mahal*. London: Thames and Hudson.

Pandey, R.N. (1985) Archeological remains of Lumbini. *Journal of the Research Center for Nepal Studies: Contributions to Nepalese Studies*, 12(3): 51–62.

Richter, L.K. (1989) *The Politics of Tourism in Asia*. Honolulu, HI: University of Hawaii Press.

—— (1999) After political turmoil: the lessons of rebuilding tourism in three Asian countries. *Journal of Travel Research*, 38: 41–45.

SAARC (2005) SAARC regional profile 2005: poverty reduction in South Asia through productive employment. Available from http://www.saarc-sec.org/data/pubs/rpp2005/pages/frameset-2.htm (accessed April 15, 2008).

—— (2008a) SAARC summit. Available from http://www.saarctourism.org/saarc-summit.html (accessed April 16, 2008).

—— (2008b) South Asian Association for Regional Cooperation. Available from http://www.saarc-sec.org/main.php (accessed April 12, 2008).

Shareef, R. and McAleer, M. (2007) Modelling the uncertainty in monthly international tourist arrivals to the Maldives. *Tourism Management*, 28: 23–45.

Sharpley, R. and Sundaram, P. (2005) Tourism: a sacred journey: the case of Ashram Tourism, India. *International Journal of Tourism Research*, 7(3): 161–71.

Singh, R.P.B. (2006) Pilgrimage in Hinduism: historical context and modern perspectives. In D.J. Timothy and D.H. Olsen (eds), *Tourism, Religion and Spiritual Journeys*, pp. 220–36. London: Routledge.

Singh, S. (2004) Religion, heritage and travel: case references from the Indian Himalayas. *Current Issues in Tourism*, 7(1): 44–65.

Tambiah, S.J. (1996) *Leveling Crowds: Ethnonationalist Conflicts and Collective Violence in South Asia*. Berkeley, CA: University of California Press.

The World Factbook (2008) Afghanistan. Available from https://www.cia.gov/library/publications/the-world-factbook/geos/af.html (accessed June 26, 2008).

Timothy, D.J. (2003) Supranationalist alliances and tourism: insights from ASEAN and SAARC. *Current Issues in Tourism*, 6(3): 250–66.

UNDP (2006) *Asia-Pacific Human Development Reports 2006*. Colombo: UNDP.

UNESCO (2008) The World Heritage Convention. Available from http://whc.unesco.org/en/convention/ (accessed Jan 10, 2008).

UNESCO World Heritage Centre (2007) World Heritage List. Available from http://whc.unesco.org/en/list (accessed August 5, 2007).

UNWTO (2007) *Report of the World Tourism Organization to the United Nations Secretary-General in Preparation for the High Level Meeting on the Mid-term*

Comprehensive Global Review of the Programme of Action for the Least Developed Countries for the Decade 2001–2010. Madrid: World Tourism Organization.

US Department of State (2008) Travel warning. Available from http://travel.state.gov/travel/cis_pa_tw/tw/tw_927.html (accessed June 24, 2008).

World Bank (2000) *The World Bank Annual Report 2000*. Washington, DC: World Bank.

—— (2006) *The World Bank Annual Report 2006*. Washington, DC: World Bank.

Zurick, D.N. (2007) Gross national happiness and environmental status in Bhutan. *The Geographical Review*, 96(4): 657–81.

9 Heritage tourism in Southwest Asia and North Africa

Contested pasts and veiled realities

Dallen J. Timothy and Rami F. Daher

Introduction

This chapter focuses on heritage tourism issues in Southwest Asia and North Africa, which includes Turkey and sometimes Azerbaijan and Afghanistan because of their cultural and religious connections to the rest of the region (Lew *et al.* 2008). The region is also often referred to as the Middle East and North Africa and, for expediency, both these designations will be used interchangeably in this chapter. For the purposes of this chapter, Central Asia and the Caucasus region are not considered but, because of their cultural and historical connections, Turkey and Iran are. The two most notable geographical elements that make this a unified region are the dominant religion (Islam) and the arid and semi-arid physical environment. Secondary variables are economic dependence on oil and the Arabic language, although there clearly are exceptions to this, as several of the states in the region have no or little petroleum resources, while Turkey, Iran, and Israel have national languages besides Arabic. Many of the countries in this region are less developed, although by most global standards several (e.g., Turkey, Lebanon, Israel, Saudi Arabia, Iran, Bahrain, Qatar, and the United Arab Emirates (UAE)) would fall into the category of developed, industrialized, or newly industrialized nations (Sönmez 2001).

North Africa and Southwest Asia is blessed with a rich and varied array of tangible and intangible culture, which gives the region one of the most bountiful resource bases for heritage tourism in the entire world. Many observers have noted this in a variety of contexts and hinted at the huge latent potential for cultural heritage tourism to develop more than it already has (Alipour and Heydari 2005; Alizadeh and Habibi 2008; Burns and Cooper 1997; Daher 2005, 2007b; Hang and Kong 2001; McGahey 2006; O'Gorman *et al.* 2007; Ouerfelli 2008; Richards 2007; Smith 2003; Tosun *et al.* 2003; Yarcan and Inelmen 2006; Zaiane 2006). Unfortunately, however, Southwest Asia and North Africa have a reputation of being dangerous destinations to visit (Alhemoud and Armstrong 1996; Gelbman 2008; Issa and Altinay 2006; Sönmez 2001)—a stereotyped image often erroneously fueled by foreign media reports. Nonetheless, the terrorist attacks of September 11, 2001, had devastating effects on tourism in the region. In addition, recent civil wars in Lebanon and Algeria; the ongoing war in Iraq; tensions between neighboring Iraq and Iran; the Iraqi invasion of Kuwait in 1990; current hostilities between Syria and Lebanon; Turkish incursions into Iraqi Kurdistan; contemporary terrorist attacks and tourist kidnappings in Yemen, Egypt, Algeria, Saudi Arabia, and Iraq; the long-term clash between Israel and its Arab neighbors; George Bush's extreme labeling of Syria and Iran as "axes of evil" (whether or not they really are) and state sponsors of terrorism, even though the US has made recent agreements with Syria regarding tensions in Lebanon; the enduring Palestinian struggle for an independent homeland; and many other current and recent events have placed this region above all else in the global media as a hotbed for conflict.

These issues, coupled with a general anti-Western sentiment fueled by foreign policies in the US (pro-Israel) and other Western states that favor certain ethnic and religious groups over others, a lack of positive and welcoming counterpromotional efforts, and much of the region's environment of poverty, have led to a failure to develop heritage tourism "analogous to its immensely rich and diverse natural, cultural and historical resources and attractions" (Sönmez 2001: 129). Clearly, there are exceptions to this generalization, such as the successes experienced in the realm of heritage tourism in Jordan, Egypt, Turkey, Tunisia, and Lebanon. The region's negative image, much of it perpetuated by Western foreign policies and media partiality, is unfortunate and has suppressed the development of tourism in a region that otherwise has a great deal to offer foreign visitors. Nearly all tourists to the region find the people to be extremely friendly and hospitable, and the individual countries to be welcoming to foreign tourists and tourism-related investments (Noack 2007; Schneider and Sönmez 1999).

Concomitantly, because of the political tensions and anti-Western views, many Western nations have issued travel advisories for several countries in the region. For example, as of July 30, 2008, the US government had posted travel advisories against visiting eleven of the twenty-one countries in the region for reasons such as kidnappings, terrorist threats, war, violence, anti-American demonstrations, embassy attacks, random arrests, and general security threats. At the same time, the Australian government had posted travel warnings regarding all countries in the region.

Notwithstanding these calamitous conditions, tourism is doing rather well in the region as a whole, better in some countries than in others. For instance, tourism was an important part of the economy of Lebanon before its civil war. The war caused tourism to plunge but, once it ended in 1990, tourism once again became important to the Lebanese economy and saw considerable recovery (Butler and Hajar 2005; Issa and Altinay 2006; Lew *et al.* 2008), although most of its tourism industries are supported by intraregional travel rather than by visitors from Europe, North America, Asia, and the Pacific (Daher 2007a; Richards 2007; Timothy and Iverson 2006). Egypt, Turkey, Tunisia, Morocco, the UAE, Oman, and Jordan have faired quite well through the political downturns. Israel's primary international market is religious tourists, who have a tendency to be less concerned with security warnings compared with other types of tourists, so Israel too has seen relatively successful growth, in spite of notable ebbs and flows in demand.

In the face of a potential decline in oil reserves in the next generation or two, several of the Gulf States are attempting to develop tourism fairly quickly. The UAE, particularly in Dubai, is probably the best example of a Gulf country with a rapid tourism development program that focuses on mass-produced, commercial, high-spend, and luxury-oriented tourism (Lew *et al.* 2008; Robatham 2005). Natural heritage, most notably desert landscapes, is the main tourism resource in other countries in the region, such as Bahrain, Qatar, Kuwait, and Oman. However, several of these countries have begun to realize the importance of their cultural heritage and have started to redirect tourism development and marketing efforts to include more emphasis on cultural heritage (Al-Azri and Morrison 2006; Gugolz 1996; Mershen 2007; Soper 2008). In Oman, for example, which is perhaps best known for desert safaris, 1994 was designated the "Year of National Heritage," reflecting new laws enacted to protect living culture and built heritage (Gugolz 1996), and tour operators there have begun involving the Bedouin nomads to enhance the heritage tourism product (Mershen 2007; Winckler 2007). Similar heritage-oriented trends are occurring in North Africa, particularly in countries such as Tunisia and Libya, whose tourism sectors have largely been based on desert safaris and beach resorts (Kohl 2006; Zaiane 2006).

The sections that follow highlight the various issues, patterns, trends, and challenges extant in Southwest Asia and North Africa today. These include issues related to pilgrimage, war and conflict, successive empires, indigenous people, and conservation challenges.

Pilgrimage

Pilgrimage refers to travel by religious adherents or spirituality seekers to places deemed sacred by the socio-religious groups to which they belong. In undertaking these kinds of journeys, pilgrims travel to sacrosanct locations to get closer to God and gain divine favor, to receive answers to prayers, to supplicate God for desired blessings, to be healed, to mingle with co-religionists, or to undergo spiritual experiences that will strengthen their faith. With only a few exceptions, most religions of the world either encourage or require this form of religious travel. On the whole, pilgrimage, or religious tourism, is extremely pervasive and socio-economically important throughout the broader Middle East region (Collins-Kreiner and Gatrell 2006; Shackley 2001), probably more so than anywhere else in the world.

As noted in Chapter 1, pilgrimage is a clear subtype of heritage tourism from at least three perspectives. First, the sites visited are heritage places. Most of these include churches, mosques, synagogues, and temples, although they also include cemeteries/graves, sacred caves, mountains, and sometimes indigenous archeological sites. These typically have different meanings for people who visit based on their own religious commitment to the place versus ordinary tourists (i.e., non-pilgrims) who visit because the place is interesting and has considerable universal value. The second perspective deals more particularly with pilgrimage routes. These, too, have become considerable heritage because, through time, by virtue of the historical practices and pathways associated with them, they have in themselves become heritage resources. The final perspective is that devout religious tourists travel as a religious practice, or to satisfy religious requirements. In this sense, the trip itself and the worship being done become intangible heritage—something of a religious or spiritual nature inherited from the past that is utilized today.

Southwest Asia is home to lands and heritage places considered holy by approximately 53 percent of the world's total population—Muslims, Christians, Jews, and various others. It is doubtful whether any other region of the world has so many designated sacred sites in relation to its size and population. Mecca (Makkah), Saudi Arabia, is one of the best known pilgrimage sites and the holiest space for all Muslims. Each year, during the month of *Zul-Hijja*, approximately two million Muslims from all parts of the earth gather together in Mecca, where they perform sacred rituals inside and outside *al-masjid al-haram* at various locations in and near Mecca. This pilgrimage (the *Hajj*) is obligatory for all Muslims inasmuch as they are financially or physically able to do it (Timothy and Iverson 2006).

The *Hajj* is considered one of the largest tourist movements and events in the world (Ahmed 1992; Aziz 2001; Burns 2007) and resembles other forms of cultural tourism from several perspectives. First, the pilgrimage employs many people directly and indirectly in the vicinities of Mecca and Medina as guides, food providers, transportation providers, accommodations workers, and other service-related jobs. Second, the *Hajj* has become more mainstream and modernized to cater to the demands of an ever more sophisticated cohort of global travelers. Air-conditioned buses, luxury hotels, and guided *Hajj* tours are now available to pilgrims who can afford them (Ahmed 1992; Delaney 1990; Timothy and Iverson 2006). For many less-affluent pilgrims, who cannot afford the expenses associated with visiting Mecca, commercial opportunities arise for them to bring traditional goods and handmade craft items to sell along the way to Mecca or once they arrive in Mecca. This helps thousands of pilgrims each year alleviate the costs of undertaking the trip to Saudi Arabia (Timothy and Iverson 2006). These commercial efforts generate interesting cultural exchanges and festivals (*Hajj* fairs) associated with the *Hajj* in different locations along travel routes.

In addition to the *Hajj*, Muslims embark on several other types of religiously motivated trips (*ziarat*) as well. These tend to focus on shrines, tombs, mosques, and other heritage sites associated with the Prophet Mohammed and famous saints, imams, and martyrs (Bhardwaj 1998). *Umrah* is another type of pilgrimage that resembles the *Hajj* and takes place in Mecca, except that it is typically shorter in duration, does not involve as many rituals, and can be done at any time during the year. These pilgrimages are not manda-tory in Islam, but are encouraged as a way of receiving blessings in one's life and demonstrating devotion to God. Trips to the holy cities of Medina and Jerusalem are considered very auspicious *ziarat*s, while visits to Islamic locations in other parts of the world (e.g., Europe, North America, Asia, Africa, etc.) are also conducive to receiving blessings and making prayers more effectual (Aziz 2001; Bhardwaj 1998; Timothy and Iverson 2006). There are many *ziarat* destinations throughout the entire Middle East, including many Shiite-specific sites in Iran, Jordan, Syria, and Iraq, which draw many pilgrims each year (Alipour and Heydari 2005; O'Gorman *et al.* 2007). In Jordan, for example, there are several shrines and graves associated with friends of the Prophet Mohammed, such as *Jafar al Tayyar*, which draws many Shiite Muslims from Iran and Iraq on religious pilgrimages.

Judaism does not require or recognize formal pilgrimage, but scores of Jews from the diaspora travel to Israel to worship at the Western Wall and to visit their ancestral Holy Land. In addition, Israelis sometimes travel to other parts of the region to visit lands inhabited by other large Jewish populations. In addition, there are many Jewish holy sites throughout the region, including Egypt, Iran, Iraq, Morocco, and Palestine (Cohen Ioannides and Ioannides 2006; Collins-Kreiner and Olsen 2004), although travel to some of these locations today is very difficult for Israelis and other Jews. For instance, visits to the Palestinian Territories by Israeli Jews is presently strictly controlled or forbidden by the Israeli government. Jewish tourists from abroad commonly visit the Holy Land for these historical reasons and to celebrate bar/bat mitzvahs—all of which Cohen Ioannides and Ioannides (2006) term "pilgrimages of nostalgia."

Finally, various sites in the Middle East are especially sacred to Christians as the locations where Jesus was born, lived, performed miracles, conducted his ministry, and died and was resurrected (Collins-Kreiner *et al.* 2006; Olsen 2006; Poria *et al.* 2007). In addition to Israel, many organized Christian tours to the Holy Land take in the less-developed countries of Jordan, Egypt, Turkey, and Palestine, because of these countries' important role in the lives and events of ancient prophets, whom Christians also revere. The Jordan River, the tradi-tional site of Jesus' baptism, for example, is visited by hundreds of thousands of tourists each year on the Jordanian side of the border; the popularity of the site increased with the visit of Pope John Paul II in March 2000, which many Roman Catholics interpreted as validating the authenticity of the site.

War and conflict

As already noted, Southwest Asia and North Africa has an unfortunate dis-tinction of being one of the most conflict-ridden parts of the world, largely based on multiple heritages. At the root of this situation is European coloni-alism, which took hold following the early twentieth-century overthrow of the Ottoman Empire and lasted until the mid-1900s. In common with many colonial areas of the world, during British and French control, the region was divided without much socio-spatial reasoning for administrative convenience and, perhaps more importantly, as many people believe, to divide and fragment a fairly unified part of the world. Thus, a region that functioned essentially as a single state under Ottoman control was carved up between colonial powers, often without an obvious rationale, and the foreign notion of nation-state was imposed on people who had not theretofore been used to it. When independence was granted for the Levant states and several North African countries in the mid-twentieth century, and when Zionist incursion in Palestine resulted in the displacement of hundreds of thousands of Palestinians, many problems, including border conflicts and secessionist movements, ensued.

Political turmoil and war directly and indirectly affect heritage and tourism in a variety of ways (Timothy and Boyd 2003). Direct effects include targeting artifacts and sites that are considered holy by opposing parties, or "heritage as target." Examples include the civil wars in Lebanon and Algeria, and the current crisis in Iraq. In all these cases, opposing sides often targeted holy sites and other heritage places deemed important by the adversary (Naccache 1998).

A second impact is heritage as an innocent casualty. While heritage sites may not be targeted directly for destruction, they are often damaged or destroyed because of their proximity to target spots. Clear examples of this exist in Iraq, where the US-led invasion destroyed many historic buildings and allowed for extensive looting of antiquities from the Iraq Museum and other locations (Bogdanos 2005).

War affects heritage tourism by decreasing arrivals and restricting access to historic places. Tourism is highly versatile, and even rumors of wars and political discontent will send tourist arrivals plummeting, even in places not directly involved in the conflict. Travel warnings are often issued by states in major potential market regions, adding an additional layer to already fragile tourism industries. Tourism clearly cannot thrive in these types of situations. Battles destroy tourism infrastructure, and heritage itself is often neglected in state budgets in favor of defense spending. Likewise, during times of crisis, tourist spending declines, resulting in even smaller budgets to devote to tourism development and heritage management. In many ways, war and conflict are enormous barriers to the conservation of artifacts and places and an obvious deterrent to tourism (Bogdanos 2005; Timothy and Boyd 2003).

In a few cases in the region, war and its remnants have become heritage attractions. In 2006, the Israeli military invaded southern Lebanon and targeted towns and villages and the southern part of the capital, Beirut, destroying much of that city's Dahiyah district. Following the conflict, reconstruction work began in the southern villages and in Beirut, bringing many volunteer non-governmental organization (NGO) workers, students, and academics to inspect the damage and assist in the reconstruction efforts. Thus, a form of volunteer tourism developed from a heritage of conflict in Lebanon. In addition, the southern suburbs of Beirut became a spectacle for Lebanese visitors and foreign tourists, particularly from other Arab states, who frequented these locations after the war to see the damage inflicted by the invasion.

In Iraq, adventuresome tourists can visit Baghdad and historic locations in the north (Lew *et al.* 2008; McGahey 2006). Borders and the landscapes of conflict that accompany them in the Middle East may be seen as heritage political landscapes that attract tourists. This is particularly the case in Israel, where tourists gather to look over borders to see what lies on the other side in Lebanon and Syria (Gelbman 2008). By the same token, a popular activity among Palestinians and other Arabs is to look over into Palestine from famous sites in Jordan (e.g., Mkies) and Syria (e.g., Golan Heights) with a longing for a return to Palestine.

Similarly, the security landscapes associated with the region receive considerable tourist attention. Another perspective related to conflict as attraction is the notion of "political tourists" who visit the Middle East either because they are simply curious about existing tensions or because they support Israel and its cause or the Palestinians and their cause (Brin 2006). Most of these "solidarity tourists" in support of Israel are Jews from the diaspora and evangelical Christians from Europe and North America who back the cause of Zionism. Pro-Palestinian solidarity tourists are not as plentiful, largely because of Israeli prohibitions, but they do exist. These people include academic solidarity groups, social justice groups, diplomats, and journalists, who usually participate in Palestinian tours of the West Bank, Gaza, and Jerusalem. Tours carry visitors through East Jerusalem and refugee camps to illustrate injustices committed by Israel; the new security wall features prominently in these circuits (Brin 2006; Timothy and Emmett forthcoming).

Related to conflict is the idea of contested heritage and place. Jerusalem, perhaps more than any other city in the world, is hotly contested between Jews, Muslims, and Christians, and even among various sects within Christianity. There are many examples in the Middle East of this kind of contestation, although most are not as remarkable. Each religion claims overlapping spaces and that its own version of the past is correct. This often leads to skirmishes in the city and has pitted many faiths against each other. The Palestinians and the Israelis, for instance, both claim Jerusalem to be their eternal and rightful capital; the city is at the core of the conflict, particularly the Temple Mount, or *Haram al-Sharif* (Timothy and Emmett forthcoming).

Layers of archeology and a succession of empires

One of the most impressive perspectives of heritage in this region is the succession of empires that lay atop one another. For instance, in Jerusalem, the present street level is not the street level of 2,000 or 3,000 years ago; the Roman period of the city lies some 10–11 meters below the present surface. This is common throughout the region, as one successive ruling power crushed the previous one and rebuilt upon it. This succession of empires and outside rulers has created a sizeable and sometimes confusing assortment of archeological periods, artifacts, and sites. Nearly all the countries of Southwest Asia and North Africa have substantial archeological remains and ancient monuments that either already appeal to tourists or have the potential to do so. Egypt, for example, ranks as one of the most desired destinations in the world, in the company of Italy, Greece, and Turkey, for people who wish to experience ancient artifacts and archeological ruins (Hang and Kong 2001).

Various parts of Southwest Asia have a long history of being ruled by outsiders (including the British and French), although many of the ruling powers have been from within the region (e.g., Egyptians, Ottomans, Persians, Arab Muslims, and Assyrians). As noted previously, many of the current states in the region only became independent or were formed as modern nation-states during the twentieth century. Sites and artifacts associated with each of these periods and many more are scattered throughout the region and are important tourism resources (Daher 2007a). In addition to the physical manifestations of these successive empires, something more influential is the socio-cultural and socio-psychological implications of this checkered political past—a collective identity crisis.

In other parts of the world that were ruled for centuries and millennia by their own people, there is a firmer identity. What it means to be Swedish, Portuguese, or Thai, for example, is fairly well established and a sense of nationhood well grounded. With few opportunities to govern themselves and lay foundations for centuries-old national and regional identities, much of North Africa and Southwest Asia is undergoing an identity crisis of sorts, which plays out significantly in the selection, conservation, and interpretation of heritage and lies at the root of so much conflict in the region. Exacerbating this problem is the pattern seen in other parts of the world (e.g., Africa, Asia, and Latin America), where the European colonial powers took it upon themselves to erect state borders for their own administrative convenience without taking into account ethnic, religious, or linguistic divisions. Thus, modern state borders are rarely congruous with national characteristics or ethnic identity.

Because of this colonial interference, even since the collapse of the Ottoman Empire in the early 1900s, few of the states in the region have been able to create and maintain a national identity that is representative and inclusive (Daher 2007a; Kumaraswamy 2006; Telhami and Barnett 2002). Many examples of this problem exist, such as that among the numerous Palestinian Arabs living inside Israel, who are also citizens of Israel. They empathize with their Palestinian co-nationals and disagree with many of Israel's policies, but they also enjoy the benefits of living in a developed country with all that it entails. From the perspective of other Arabs, being Israeli is incompatible with being a Palestinian or an Arab, which deepens the identity predicament even further (Baumeister *et al.* 1985; Rouhana and Ghanem 1998). Other instances like this exist in the region where ethnic minorities face challenges assimilating and finding an equal footing in the state where they reside. Examples include the Kurds in Turkey and Iraq, the Azeris and other Turkic peoples of Iran, the Coptic Christians in Egypt, and Armenians in Iran, Syria, Iraq, and Turkey.

Another character crisis exists in Turkey. While Turkish youth today identify well with the Turkish state, it was more difficult for their grandparents and great-grandparents to come to terms with a Turkish identity (rather than a Turkic identity), as Turkey was a relatively new country, formed from the rubble of the Ottoman Empire (Özdoğan 1998). Based largely on the effects of a Turkic identity (which originates in Central Asia), the prominent religion (secular Islam), and other sundry variables, Turkish national identity is today plagued with questions of where it belongs in the world. Is Turkey European, Asian, or Middle Eastern? Clearly this matters in Turkey in creating a national identity, in generating allegiance to the state, and in bringing to fruition its desires for accession to the European Union (Robins 1996). Thus, national heritage in the Middle East is difficult to define, and not all citizens are eager to support what others wish to consider their national past.

Indigenous peoples

There are many different indigenous peoples of Southwest Asia and North Africa, but the group receiving most attention these days in the tourism research literature is the desert nomads (Al-Oun and Al-Homoud 2008; Aziz 2001; Bos-Seldenthuis 2007; Homa 2007). Such a view of indigenousness in the Middle East, given the region's much larger array of ethnic groups, is a highly Orientalist perspective (Said 1979), and is indicative of the stereotypes perpetuated in the media and in some marketing campaigns where tourists to the region still imagine themselves "coming to a cultural landscape that ha[s] not changed since antiquity, or since ancient Biblical and Byzantine times" and where they romanticize a "non-changed village life in Palestine and Syria" (Daher 2007a: 7). Nonetheless, much of the tourist worldview is Orientalist in nature and continues to focus on the noble heritage of the nomads of the Middle East.

In common with many other nomadic societies of the world (e.g., the Mongols), there are few tangible elements of nomadic cultures remaining in this region. As already noted, most heritage sites and artifacts utilized as tourism resources are vestiges of the colonial and imperial periods (Daher 2007a, 2007b; Porter and Salazar 2005), overshadowing what relatively little remains from prehistoric eras (Finegan 1979; Gopher *et al.* 2002). As a result, indigenous heritage in North Africa, the Arabian Peninsula, and the Middle East emphasizes the intangible patrimony of the region's Bedouins and other nomadic peoples, who comprise some 10 percent of the realm's total population. As Bos-Seldenthuis (2007: 32) notes, "The material culture of the Ababda nomads is meager, as is often the case with migrating people, while in contrast their oral traditions and other expressions of intangible culture are vast." While a few items of material culture are important in the heritage of nomads, such as their tents, food, and dress, their non-material culture, of which they are especially proud, includes their knowledge and use of the desert environment, hunting prowess, ability to raise animals in arid regions, knowledge of the healing properties of certain plants, religious beliefs, languages, music, sport, craftsmanship, and inherent hospitality (Bos-Seldenthuis 2007).

These items, together with the very idea of meeting nomadic peoples, creates a tourist appeal unique to this region. The Bedouins, Tuaregs, and Ababda nomads of Syria, Jordan, Saudi Arabia, Oman, UAE, Egypt, and Libya are becoming more involved in the tourism sector as guides, cooks, and souvenir vendors (Bos-Seldenthuis 2007; Kohl 2006). In Jordan, several projects are under way where tourists visit and stay overnight in the tent homes and encampments of the Bedouins, sharing food, music, and animal care (Al-Oun and Al-Homoud 2008; Lew *et al.* 2008).

The commodification of Bedouin culture is becoming more widespread, and increasing numbers of nomads are becoming involved in the production of tourism (Homa 2007). Shoup (1985: 283) noted how one hotel in Jordan purchased a large Bedouin tent and pays Bedouins to staff the tent offering indigenous hospitality to guests "who wish a 'real taste' of Bedouin life." Another implication is forced relocations. In Jordan, for example, when Petra was designated a national park and tier-one tourist attraction, the Bidul people experienced tremendous trauma when they were forced to move from Petra to a manufactured village with concrete homes and government-provided electricity (Shoup 1985). Throughout the region, grazing lands are also being abandoned through voluntary relocations as nomads move to cities in search of work in tourism and tourism-related construction (Bos-Seldenthuis 2007).

Historic cities and the region's urban heritage

While the idealized lifestyles of the nomad appeal to many foreign tourists, one cannot ignore the importance of historic villages and cities in the region. Although the desert nomads have little by way of material culture, there is a vibrant and remarkable architectural heritage in the region's villages and cities (Daher 1999). Many cities in the Middle East and North Africa are world renowned for their fine heritage composition and heralded as well-preserved examples of urban development, remnants of ancient empires, and vestiges of impressive Islamic architecture. Some of the larger cities include Cairo, Damascus, Tunis, Jerusalem, Sana'a, Istanbul, Tripoli, and Baghdad, with many secondary cities and small towns being of equal architectural and archeological value: Bam, Iran; Aanjar, Baalbeck, Tyre, and Byblos in Lebanon; Meknes and Marrakesh, Morocco; Aleppo, Syria; Salt, Jordan; and Shibam, Yemen, to name but a few. Many of these have been recognized as having universal cultural value and have been placed on UNESCO's World Heritage List (see Figure 9.1).

Many *medina*s, or Islamic cores of cities in North Africa, have become a driving force for the development of urban heritage tourism. Cities such as Tunis, Casablanca, Fes, Tangier, and Marrakesh have developed much of their heritage tourism product around their famous *medina*s, with their associated mosques, markets, palaces, narrow alleyways, and fountains (Orbaşli 2000; Rghei and Nelson 1994; Serageldin and El-Sadek 1982). The combination of historic buildings, market places, religious sites, and cafés and restaurants creates an ambience and appeal that is unique to the region and to Arab Muslim cities in general.

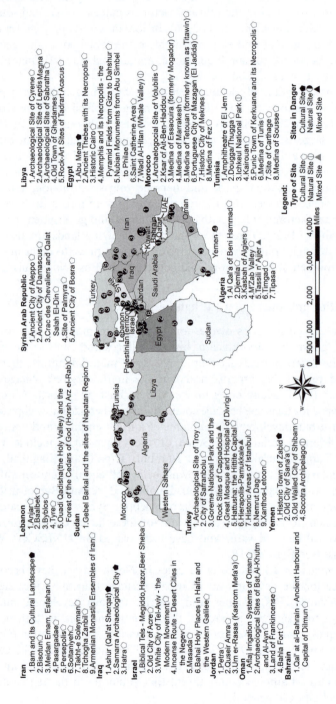

Iran
1. Bam and its Cultural Landscape ●
2. Bisotun ○
3. Meidan Emam, Esfahan ○
4. Pasargadae ○
5. Persepolis ○
6. Soltaniyeh ○
7. Takht-e Soleyman ○
8. Tchogha Zanbil ○
9. Armenian Monastic Ensembles of Iran ○

Iraq
1. Ashur (Qal'at Sherqat) ●
2. Samarra Archaeological City ●
3. Hatra ○

Israel
1. Biblical Tels - Megiddo,Hazor,Beer Sheba ○
2. Old City of Acre ○
3. White City of Tel-Aviv - the Modern Movement ○
4. Incense Route - Desert Cities in the Negev ○
5. Masada ○
6. Bahai Holy Places in Haifa and the Western Galilee ○

Jordan
1. Petra ○
2. Quseir Amra ○
3. Um er-Rasas (Kastrom Mefa'a) ○

Oman
1. Aflaj Irrigation Systems of Oman ○
2. Archaeological Sites of Bat,Al-Khutm and Al-Ayn ○
3. Land of Frankincense ○
4. Bahla Fort ○

Bahrain
1. Qal' at al-Bahrain - Ancient Harbour and Capital of Dilmun ○

Lebanon
1. Anjar ○
2. Baalbek ○
3. Byblos ○
4. Tyre ○
5. Ouadi Qadisha(the Holy Valley) and the Forest of the Cedars of God (Horsh Arz el-Rab) ○

Sudan
1. Gebel Barkal and the sites of Napatan Region ○

Syrian Arab Republic
1. Ancient City of Aleppo ○
2. Ancient City of Damascus ○
3. Crac des Chevaliers and Qalat Salah El-Din ○
4. Site of Palmyra ○
5. Ancient City of Bosra ○

Turkey
1. Archaeological Site of Troy ○
2. City of Safranbolu ○
3. Goreme National Park and the Rock Sites of Cappadocia ▲
4. Great Mosque and Hospital of Divrigi ○
5. Hattusha: the Hittite Capital ○
6. Hierapolis-Pamukkale ▲
7. Historic Areas of Istanbul ○
8. Nemrut Dag ○
9. Xanthos-Letoon ○

Yemen
1. Historic Town of Zabid ●
2. Old City of Sana'a ○
3. Old Walled City of Shibam ⊕
4. Socotra Archipelago ○

Algeria
1. Al Qal'a of Beni Hammad ○
2. Djemila ○
3. Kasbah of Algiers ○
4. M'Zab Valley ○
5. Tassili n' Ajjer ▲
6. Timgad ○
7. Tipasa ○

Libya
1. Archaeological Site of Cyrene ○
2. Archaeological Site of Leptis Magna ○
3. Archaeological Site of Sabratha ○
4. Old Town of Ghadames ○
5. Rock-Art Sites of Tadrart Acacus ○

Egypt
1. Abu Mena ●
2. Ancient Thebes with its Necropolis ○
3. Historic Cairo ○
4. Memphis and its Necropolis – the Pyramid Fields from Giza to Dahshur ○
5. Nubian Monuments from Abu Simbel to Philae ○
6. Saint Catherine Area ○
7. Wadi Al-Hitan (Whale Valley) ○

Morocco
1. Archaelogical Site of Volubilis ○
2. Ksar of Ait-Ben-Haddou ○
3. Medina of Essaouira (formerly Mogador) ○
4. Medina of Marrakesh ○
5. Medina of Tetouan (formerly known as Titawin) ○
6. Portuguese City of Mazagan (El Jadida) ○
7. Historic City of Meknes ○
8. Medina of Fez ○

Tunisia
1. Amphitheatre of El Jem ○
2. Dougga/Thugga ○
3. Ichkeul National Park ⊕
4. Kairouan ○
5. Punic Town of Kerkuane and its Necropolis ○
6. Medina of Tunis ○
7. Site of Carthage ○
8. Medina of Sousse ○

Legend:

Type of Site		Sites in Danger	
Cultural Site	○	Cultural Site	●
Natural Site	⊕	Natural Site	●
Mixed Site	▲	Mixed Site	▲

0 500 1,000 2,000 3,000 4,000
Miles

Figure 9.1 Developing countries and World Heritage Sites in the Middle East and North Africa

These historic cities face problems similar to those around the world in that crowded conditions, urban development, and modernization have in some cases overpowered efforts to preserve the past (Ibrahim 2001; Shechter and Yacobi 2005). These human-induced issues, together with natural pressures, have caused a few of the region's most fascinating ancient cities to be placed on UNESCO's list of World Heritage in Danger: the historic town of Zabid, Yemen; the Old City of Jerusalem; Bam, Iran. Nonetheless, overall, the ancient cities are well preserved and will no doubt continue to be important heritage tourism destinations far into the future, just as they have been for centuries in the past.

Heritage challenges

Aside from direct hits via war, political turmoil, and the heritage contestation already noted, there are several other problems facing the conservation of heritage in the Middle East. One of them is war related. During the first Gulf War of 1990–91, Iraqi soldiers withdrawing from Kuwait set fire to many of that country's oil wells. This had a devastating environmental effect, not least upon the built heritage of the region. The effect was especially profound in Iran, Iraq, and Kuwait, where gases, sulfuric acid, soot, and polycyclic aromatic hydrocarbons were believed to have covered and caused the deterioration of many of the region's oldest monuments and heritage structures. Fortunately, however, one study suggests that the erosive elements did not proliferate very far (Bonazza *et al.* 2007).

Pollution-related problems exist throughout the entire realm, especially in highly urbanized areas such as Cairo, one of the most polluted cities in the world (Smith 2003). The Pyramids at Giza have long been under threat of urban encroachment and the accompanying effects of human effluent, litter, and smog (Hang and Kong 2001; Timothy 1994), which deteriorate delicate construction materials and create health hazards for tourists and residents.

One of the most common challenges is overuse of historic sites. Despite the sharp fluctuations in regional arrivals, several sites are heavily visited year round almost regardless of the political climate. Perhaps the best example is the Pyramids in Cairo. Masses of tourists climbing on and in the pyramids have been causing damage for years with added humidity in the tombs and wear and tear on the structure itself (Mitchell 2001; Timothy 1994, 1999). The Pyramids' importance as a global tourist attraction renders them open to this kind of overuse. Fortunately, however, owing to foreign pressure, not least by UNESCO, the site is now managed better and in a more environmentally sustainable manner with access restricted and clean-up projects under way (Hang and Kong 2001; Smith 2003). In large part because of the pressures of mass tourism, several important World Heritage Sites in the region feature prominently on UNESCO's list of World Heritage in Danger. In fact eight of the thirty sites on the list are located in this region (Figure 9.1). The Pyramids of Giza are no longer on the danger list, but were placed there several years ago to pressure the Egyptian government into improving their management and conservation efforts. The Old City of Jerusalem is also on the list, being cited as threatened by mass tourism, urban development, and lack of conservation. Other reasons listed for the eight sites in danger include bad planning (irrigation, flooding, and damming), dubious conservation efforts, political tension, lack of maintenance, and urban encroachment (UNESCO 2008).

Quantities of the famed Cedars of Lebanon—unique cedar trees made famous by their Biblical associations—have been depleted to a large extent and are now endangered because of overuse and a lack of protective measures in place during the French colonial and Ottoman periods (Abu-Izzeddin 2000; Shackley 2004). This symbol of Lebanese nationality and heritage is an important resource for tourism, and the two remaining small tracts of trees being preserved in the country have become major attractions.

Ironically, the abundance of archeological sites and other historic places may be one of the region's most significant challenges. The least developed countries of North Africa and Southwest Asia have difficulty in preserving their many heritages. Hang and Kong (2001) note that the maintenance of existing and excavated sites is already a burden on countries such as Egypt and, with increased excavations, already onerous financial and staffing burdens are made even more stressful and difficult. Some countries' antiquities ministries are already past their saturation point and pragmatically have little interest in conducting or allowing additional excavations and development projects.

With one exception, all the countries of the region are sovereign states that are able to exert control over tourism and heritage management. The exception is Palestine, which resembles a labyrinth polity of a sovereign state, a puppet state, and an occupied state. Palestine faces perhaps the most insurmountable tourism challenges of all its neighbors—troubles created by its relationship with Israel. Being at the mercy of Israel, the Palestinian Territories are having a very difficult time developing tourism industries that will provide much needed income and foreign exchange for the government and individuals.

Because tourists cannot cross directly into Palestine by land, sea, or air without transiting through Israel, Israel in effect controls visitor flows and even tourist spending in the Territories. Only Israel has the authority to issue visas for foreign tourists visiting the Holy Land; the Palestinian Authority is powerless to issue visas or control its own borders. In the words of Al-Rimmawi (2003: 78), the Israelis "execute their policies in such a way as to be the only franchised power." This manifests in several ways. First, Israel has erected the Security Wall around the Palestinian-controlled areas of the West Bank. Access through this wall to important tourist destinations on the Palestinian side, such as Bethlehem, is funneled through very few crossing points and, sometimes, when deemed necessary for security reasons, the gateways are unilaterally closed. This clearly has implications for tourism in Palestine, but also for tourism in Israel, as many Christian tours visit Bethlehem on day-trips to visit the birthplace of Jesus Christ. When the border is closed, tour guides and agencies are required to find alternative attractions and have to answer to upset tourists who desire the Bethlehem experience.

The positive economic effects are also controlled by Israel. Al-Rimmawi (2003) notes that, in the early 2000s, the Israeli authorities only allowed tourist groups to stay in Bethlehem for short periods of time, although this does not appear to be the case today. The Palestinians are concerned that very little economic impact can occur in just one or two hours. Israel typically gets most of the other revenue as well, because it controls most of the transportation, lodging (not in Palestine), food services, and other services utilized by tourists (Al-Rimmawi 2003; Clarke 2002; Handal 2006).

Conclusion

Southwest Asia and North Africa is composed of layer upon layer of archeologies, successive empires, cacophonous pasts, and modern-day groups vying for control of sacred and economic space. However, it is also home to some of the most spectacular remnants of human civilization on earth and worthy of much more scholarly attention. These conditions make this region one of the most complex in the entire world and bestow one of the richest foundations for heritage tourism. Unfortunately, political tensions and hyped-up negative exposure in the West have kept much of this potential from being realized and developing into a tourism sector that could benefit many of the less-developed countries of the region.

In relation to this, since the tragic events of September 11, 2001, Western tourism to the region has remained volatile. Nonetheless, many tourists from North America, Western Europe, Japan, and Australia and New Zealand are choosing to ignore many politically motivated travel warnings and are enjoying what the region has to offer. In addition, domestic and intraregional travel has accelerated, as many Arabs have elected to show solidarity with their compatriots in the Middle East. Also, because they often face various forms of prejudice and discrimination while traveling outside the region, traveling to the West has become less desirable, so that many people in Southwest Asia and North Africa have chosen to vacation in countries and regions that respect their religious mores and unique social needs (Timothy and Iverson 2006).

Irony and dichotomy are two key words in understanding heritage in this region. Wars have been and continue to be fought over heritages and homelands. Ironically, in spite of their tragic consequences, these wars and the heritages that produce them are among the most interesting and widely visited heritage attractions today. Also, travel warnings and the blind eye of the West have kept many countries at arm's length and unknown in the Western world even though they possess some of the most promising cultural heritage resources in the world. Yemen, for example, is endowed with beautiful, interesting, and rare urban architecture with which no other city in the world can compare (Burns and Cooper 1997). Yet, few people in the developed world understand the value of the nation or the potential of the hidden treasures located there. The same can be said of Lebanon, Syria, Iran, Palestine, and Libya. Despite these travel warnings and oftentimes overblown media reports, the heritage places in the Middle East and North Africa are among the most visited in the world and, according to many tourists, the benefits of visiting outweigh the potential, sometimes exaggerated, dangers.

The region is steeped in traditions that are deeply valued by its own inhabitants and by billions of people from other parts of the globe. It is looked upon as one of the hearths of human culture and innovation and one of the first areas inhabited by humans. Pilgrimage places, archeological sites, and the remains of successive empires attest to this important role. Dichotomously, it is a region of massive change—everywhere, in the wealthier oil states and the poorer agrarian states. Understanding that the world will probably become less dependent on oil in the future and that oil reserves are a finite resource, the Gulf States have begun to explore other economic options, including tourism. While much of their present efforts focus on mass and luxury tourism, many efforts are under way to broaden the base to include archeological and living heritage, as well as urban architecture and Arab city ambience. This has been instrumental in developing efforts to gentrify ancient cities and *medina* zones (Daher 1999). Likewise, the traditional peoples are becoming more involved in service industries based on foreign interest in their cultures and their need to survive in a difficult climate and economy. In facing these and other challenges, the region will carry on changing in dichotomous and ironic ways and, in so doing, will continue to create and re-create a heritage identity of its own that will continue to fascinate tourists in the foreseeable future.

References

Abu-Izzeddin, F. (2000) The protected areas project in Lebanon: conserving an ancient heritage. *Parks*, 10(1): 25–32.

Ahmed, Z.U. (1992) Islamic pilgrimage (Hajj) to Ka'aba in Makkah (Saudi Arabia): an important international tourist activity. *Journal of Tourism Studies*, 3(1): 35–43.

Al-Azri, H.I. and Morrison, A.M. (2006) Measurement of Oman's destination image in the US. *Tourism Recreation Research*, 31(2): 85–89.

Alhemoud, A.M. and Armstrong, E.G. (1996) Image of tourism attractions in Kuwait. *Journal of Travel Research*, 34(4): 76–80.

Alipour, H. and Heydari, R. (2005) Tourism revival and planning in Islamic Republic of Iran: challenges and prospects. *Anatolia*, 16(1): 39–61.

Alizadeh, H. and Habibi, K. (2008) Structural elements and the built environment. *International Journal of Environmental Research*, 2(2): 153–64.

Al-Oun, S. and Al-Homoud, M. (2008) The potential for developing community-based tourism among the Bedouins in the Badia of Jordan. *Journal of Heritage Tourism*, 3 (1): 36–54.

Al-Rimmawi, H.A. (2003) Palestinian tourism: a period of transition. *International Journal of Contemporary Hospitality Management*, 15(2): 76–85.

Aziz, H. (2001) The journey: an overview of tourism and travel in the Arab/Islamic context. In D. Harrison (ed.), *Tourism in the Less Developed World: Issues and Case Studies*, pp. 151–59. Wallingford: CABI.

Baumeister, R.F., Shapiro, J.P. and Tice, D.M. (1985) Two kinds of identity crisis. *Journal of Personality*, 53(3): 407–24.

Bhardwaj, S.M. (1998) Non-hajj pilgrimage in Islam: a neglected dimension of religious circulation. *Journal of Cultural Geography*, 17(2): 69–87.

Bogdanos, M. (2005) The casualties of war: the truth about the Iraq Museum. *American Journal of Archaeology*, 109: 477–526.

Bonazza, A., Sabbioni, C., Ghedini, N., Hermosin, B., Jurado, V., Gonzales, J.M. and Saiz-Jimenez, C. (2007) Did smoke from the Kuwait oil well fires affect Iranian archaeological heritage? *Environmental Science and Technology*, 41(7): 2378–86.

Bos-Seldenthuis, J. (2007) Life and tradition of the Ababda nomads in the Egyptian desert, the junction between intangible and tangible heritage management. *International Journal of Intangible Heritage*, 2: 31–43.

Brin, E. (2006) Politically-oriented tourism in Jerusalem. *Tourist Studies*, 6(3): 215–43.

Burns, P. (2007) From hajj to hedonism? Paradoxes of developing tourism in Saudi Arabia. In R.F. Daher (ed.), *Tourism in the Middle East: Continuity, Change and Transformation*, pp. 215–36. Clevedon: Channel View.

Burns, P. and Cooper, C. (1997) Yemen: tourism and a tribal–Marxist dichotomy. *Tourism Management*, 18(8): 555–63.

Butler, R.W. and Hajar, R. (2005) After the war: ethnic tourism to Lebanon. In G. J. Ashworth and R. Hartmann (eds), *Horror and Human Tragedy Revisited: the Management of Sites of Atrocities for Tourism*, pp. 211–23. New York: Cognizant.

Clarke, R. (2002) Self-presentation in a contested city: Palestinian and Israeli political tourism in Hebron. *Anthropology Today*, 16(5): 12–18.

Cohen Ioannides, M. and Ioannides, D. (2006) Global Jewish tourism: pilgrimages and remembrance. In D.J. Timothy and D.H. Olsen (eds), *Tourism, Religion and Spiritual Journeys*, pp. 156–71. London: Routledge.

Collins-Kreiner, N. and Gatrell, J.D. (2006) Tourism, heritage and pilgrimage: the case of Haifa's Baha'i Garden. *Journal of Heritage Tourism*, 1(1): 32–50.

Collins-Kreiner, N. and Olsen, D.H. (2004) *Selling diaspora—producing and segmenting the Jewish diaspora tourism market*. In T. Coles and D.J. Timothy (eds), *Tourism, Diasporas and Space*, pp. 279–90. London: Routledge.

Collins-Kreiner, N., Kliot, N., Mansfeld, Y. and Sagi, K. (2006) *Christian Tourism to the Holy Land: Pilgrimage During Security Crisis*. Aldershot: Ashgate.

Daher, R.F. (1999) Gentrification and the politics of power, capital, and culture in an emerging Jordanian heritage industry. *Traditional Dwellings and Settlement Review*, 10(2): 33–47.

—— (2005) Urban regeneration/heritage tourism endeavors: the case of Salt, Jordan "local actors, international donors, and the state". *International Journal of Heritage, Studies*, 11(4): 289–308.

—— (2007a) Reconceptualizing tourism in the Middle East: place, heritage, mobility and competitiveness. In R.F. Daher (ed.), *Tourism in the Middle East: Continuity, Change and Transformation*, pp. 1–69. Clevedon: Channel View.

—— (2007b) Tourism, heritage, and urban transformations in Jordan and Lebanon: emerging actors and global–local juxtapositions. In R.F. Daher (ed.), *Tourism in the Middle East: Continuity, Change and Transformation*, pp. 263–307. Clevedon: Channel View.

Delaney, C. (1990) The "hajj": sacred and secular. *American Ethnologist*, 17(3): 513–30.

Finegan, J. (1979) *Archaeological History of the Ancient Middle East*. Boulder, CO: Westview Press.

Gelbman, A. (2008) Border tourism in Israel: conflict, peace, fear and hope. *Tourism Geographies*, 10(2): 193–213.

Gopher, A., Abbo, S. and Lev-Yadun, S. (2002) The "when", the "where" and the "why" of the Neolithic revolution in the Levant. *Documenta Praehistorica*, 28: 49–61.

Gugolz, A. (1996) The protection of cultural heritage in the sultanate of Oman. *International Journal of Cultural Property*, 5: 291–309.

Handal, J. (2006) Rebuilding city identity through history: the case of Bethlehem-Palestine. In R. Zetter and G.B. Watson (eds), *Designing Sustainable Cities in the Developing World*, pp. 51–68. Aldershot: Ashgate.

Hang, L.K.P. and Kong, C. (2001) Heritage management and control: the case of Egypt. *Journal of Quality Assurance in Hospitality and Tourism*, 2(1/2): 105–17.

Homa, D. (2007) Touristic development in Sinai, Egypt: Bedouin, visitors, and government interaction. In R.F. Daher (ed.), *Tourism in the Middle East: Continuity, Change and Transformation*, pp. 237–62. Clevedon: Channel View.

Ibrahim, S.E. (2001) Addressing the social context in cultural heritage management: historic Cairo. In I. Serageldin, E. Shluger and J. Martin-Brown (eds), *Historic Cities and Sacred Sites: Cultural Roots for Urban Futures*, pp. 186–92. Washington, DC: The World Bank.

Issa, I.A. and Altinay, L. (2006) Impacts of political instability on tourism planning and development: the case of Lebanon. *Tourism Economics*, 12(3): 361–81.

Kohl, I. (2006) Von Tuareg, Toyotas und Wüsten Geschichten: Sahara-Tourismus in Libyen. *Integra*, 2: 14–17.

Kumaraswamy, P.R. (2006) Who am I? The identity crisis in the Middle East. *Middle East Review of International Affairs*, 10(1): 63–73.

Lew, A.A., Hall, C.M. and Timothy, D.J. (2008) *World Geography of Travel and Tourism: A Regional Approach*. Oxford: Butterworth Heinemann.

McGahey, S. (2006) Tourism development in Iraq: the need for support from international academia. *International Journal of Tourism Research*, 8(3): 235–39.

Mershen, B. (2007) Development of community-based tourism in Oman: challenges and opportunities. In R.F. Daher (ed.), *Tourism in the Middle East: Continuity, Change and Transformation*, pp. 188–214. Clevedon: Channel View.

Mitchell, T. (2001) Making the nation: the politics of heritage in Egypt. In N. AlSayyad (ed.), *Consuming Tradition, Manufacturing Heritage: Global Norms and Urban Forms in the Age of Tourism*, pp. 212–39. London: Routledge.

Naccache, A.F.H. (1998) Beirut's memoryside: hear no evil, see no evil. In L. Meskell (ed.), *Archaeology Under Fire: Nationalism, Politics, and Heritage in the Eastern Mediterranean and Middle East*, pp. 140–58. London: Routledge.

Noack, S. (2007) *Doing Business in Dubai and the United Arab Emirates*. Vienna: GRIN Verlag.

O'Gorman, K., McLellan, L.R. and Baum, T. (2007) Tourism in Iran: central control and indigeneity. In R. Butler and T. Hinch (eds), *Tourism and Indigenous Peoples: Issues and Implications*, pp. 251–64. Oxford: Butterworth Heinemann.

Olsen, D.H. (2006) Tourism and informal pilgrimage among Latter-day Saints. In D.J. Timothy and D.H. Olsen (eds), *Tourism, Religion and Spiritual Journeys*, pp. 254–70. London: Routledge.

Orbaşli, A. (2000) *Tourists in Historic Towns: Urban Conservation and Heritage Management*. London: E. & F.N. Spon.

Ouerfelli, C. (2008) Co-integration analysis of quarterly European tourism demand in Tunisia. *Tourism Management*, 29(1): 127–37.

Özdoğan, M. (1998) Ideology and archaeology in Turkey. In L. Meskell (ed.), *Archaeology Under Fire: Nationalism, Politics, and Heritage in the Eastern Mediterranean and Middle East*, pp. 111–23. London: Routledge.

Poria, Y., Biran, A. and Reichel, A. (2007) Different Jerusalems for different tourists: capital cities—the management of multi-heritage site cities. *Journal of Travel and Tourism Marketing*, 22(3/4): 121–38.

Porter, B.W. and Salazar, N.B. (2005) Heritage tourism, conflict, and the public interest: an introduction. *International Journal of Heritage Studies*, 11(5): 361–70.

Rghei, A.S. and Nelson, J.G. (1994) The conservation and use of the walled city of Tripoli. *Geographical Journal*, 16(2): 143–58.

Richards, G. (2007) Introduction: global trends in cultural tourism. In G. Richards (ed.), *Cultural Tourism: Global and Local Perspectives*, pp. 1–24. New York: Haworth.

Robatham, M. (2005) Creating Dubailand. *Leisure Management*, 25(5): 50–54.

Robins, K. (1996) Interrupting identities: Turkey/Europe. In S. Hall and P. Du Gay (eds), *Questions of Cultural Identity*, pp. 61–86. London: Sage.

Rouhana, N. and Ghanem, A. (1998) The crisis of minorities in ethnic states: the case of Palestinian citizens in Israel. *International Journal of Middle East Studies*, 30(3): 321–46.

Said, E.W. (1979) *Orientalism*. London: Routledge.

Schneider, I. and Sönmez, S. (1999) Exploring the touristic image of Jordan. *Tourism Management*, 20(4): 539–42.

Serageldin, I. and El-Sadek, S. (1982) *The Arab City: Its Character and Islamic Cultural Heritage*. Riyadh: The Arab Urban Development Institute.

Shackley, M. (2001) *Managing Sacred Sites*. London: Continuum.

—— (2004) Managing the cedars of Lebanon: botanical gardens or living forests? *Current Issues in Tourism*, 7(4/5): 417–25.

Shechter, R. and Yacobi, H. (2005) Cities in the Middle East: politics, representation and history. *Cities*, 22(3): 183–88.

Shoup, J. (1985) The impact of tourism on the Bedouin of Petra. *Middle East Journal*, 39(2): 277–91.

Smith, M.K. (2003) *Issues in Cultural Tourism Studies*. London: Routledge.

Sönmez, S. (2001) Tourism behind the veil of Islam: women and development in the Middle East. In Y. Apostolopoulos, S. Sönmez and D.J. Timothy (eds), *Women as Producers and Consumers of Tourism in Developing Regions*, pp. 113–42. Westport, CT: Praeger.

Soper, A. (2008) Digging into heritage tourism in Ras Al Khaimah, UAE. Paper presented at the conference "Heritage and Cultural Tourism: The Present and Future of the Past," Jerusalem, 17–19 June.

Telhami, S. and Barnett, M. (2002) *Identity and Foreign Policy in the Middle East*. Ithaca: Cornell University Press.

Timothy, D.J. (1994) Environmental impacts of heritage tourism: physical and sociocultural perspectives. *Manusia dan Lingkungan*, 4(2): 37–49.

—— (1999) Built heritage, tourism and conservation in developing countries: challenges and opportunities. *Tourism*, 4: 5–17.

Timothy, D.J. and Boyd, S.W. (2003) *Heritage Tourism*. London: Prentice Hall.

Timothy, D.J. and Emmett, C.F. (forthcoming) Jerusalem, tourism and the politics of heritage. In M. Adelman and M. Elman (eds), *Jerusalem Across Disciplines*. Bloomington, IN: Indiana University Press.

Timothy, D.J. and Iverson, T. (2006) Tourism and Islam: considerations of culture and duty. In D.J. Timothy and D.H. Olsen (eds), *Tourism, Religion and Spiritual Journeys*, pp. 186–205. London: Routledge.

Tosun, C., Timothy, D.J. and Öztürk, Y. (2003) Tourism growth, national development and regional inequality in Turkey. *Journal of Sustainable Tourism* 11(2): 31–49.

UNESCO (2008) List of World Heritage in Danger. Available from http://whc.unesco.org/en/danger/ (accessed July 30, 2008).

Winckler, O. (2007) The birth of Oman's tourism industry. *Tourism*, 55(2): 221–34.

Yarcan, & Sogenek;. and Inelmen, K. (2006) Perceived image of Turkey by US-citizen cultural tourists. *Anatolia*, 17(2): 305–13.

Zaiane, S. (2006) Heritage tourism in Tunisia: development one-way choice. *Tourism Review*, 61(3): 26–31

10 Tourism and Africa's tripartite cultural past

Victor B. Teye

Introduction

This chapter examines heritage and tourism in Africa, a region whose definition can pose a number of problems. The term subSaharan Africa is widely used to distinguish the area that largely excludes North Africa but actually includes much of the Sahara Desert itself. For example, most World Bank datasets exclude the countries of North Africa. While Africa has several islands with significant tourism industries in both the Atlantic and the Indian oceans, some refer to continental Africa, which excludes such islands as Madagascar, Seychelles, Mauritius, and Cape Verde. Another example of this problem of definition is illustrated by the United Nations World Tourism Organization (UNWTO) regions. While most North African countries are part of the Africa region, Egypt, which is actually the leading tourism destination on the continent with vast and diverse cultural and heritage tourism attractions, is part of the larger Middle East region. For the purpose of this chapter, Africa is defined as the continent and its islands, consisting mostly of the fifty-three countries that make up the membership of the African Union. While North Africa will be mentioned occasionally, the thrust of this chapter focuses on the countries south of the Sahara. The countries of Africa are listed in Table 10.1, which also shows their geographical size and population.

   Africa possesses a number of geographical attributes that constitute significant natural and cultural elements that also lay the foundation for its tourism industry. First, it occupies 18,835,221 square kilometers, making it the second largest land mass after Eurasia. Second, from the north, it stretches about 5,000 miles from Bizerte in Tunisia to Cape Agulhas in South Africa. The east–west distance is almost the same, from Dakar in Senegal to Cape Gardafui in Somalia. Third, it is the only continent positioned astride the Equator and extending almost to latitude 35 degrees south and beyond 35 degrees north of the Equator. The result of Africa's geographical location and large size is a region endowed with multiple physical tourism resources, which include a diverse array of relief forms, topography, fauna, flora, and maritime and aquatic resources (Best and de Blij 1977). Similarly, the inventory of historical and cultural resources is enormous and diverse. These include the

Table 10.1 Profile of African countries, 2006

Country	Area (000 sq km)	Population (millions)	International tourists (thousands)
Algeria	2,381.7	33.4	–
Angola	1,246.7	16.6	–
Benin	112.6	8.8	–
Botswana	581.7	1.9	–
Burkina Faso	274	14.4	–
Burundi	27.8	8.2	–
Cameroun	27.8	8.2	–
Cape Verde	4	0.5	242
Central African Republic	4	0.5	–
Chad	1,284	10.5	–
Comoros	1.9	0.6	–
Congo Republic	342	3.7	–
Democratic Republic of Congo	2,344.9	60.6	–
Djibouti	23.2	0.8	–
Egypt	1,001.5	74.2	–
Equatorial Guinea	28.1	0.5	–
Eritrea	117.6	4.7	–
Ethiopia	1,101.3	77.2	–
Gabon	267.7	1.3	–
Gambia	11.3	1.7	–
Ghana	238.5	23.0	442
Guinea	–	–	–
Guinea Bissau	36.1	1.7	–
Ivory Coast	–	–	–
Kenya	580.4	36.6	–
Lesotho	30.4	2.0	357
Liberia	111.4	3.6	–
Libya	1,759.5	6.0	–
Madagascar	1,759.5	6.0	312
Malawi	118.5	13.6	–
Mali	1,240.2	12.0	–
Mauritania	1,030.7	3.0	–
Mauritius	2.0	1.3	788
Mozambique	799.4	21	–
Namibia	824.3	2.1	–
Niger	1,267	13.7	–
Nigeria	923.8	144.7	–
Rwanda	26.3	9.5	–
Sao Tome and Principe	1	0.2	–
Senegal	196.7	12.1	–
Seychelles	0.5	0.1	141
Somalia	637.7	8.5	–
South Africa	1,219.1	47.4	8,396
Sudan	2,505.6	37.7	328
Swaziland	17.4	1.1	873
Tanzania	947.3	39.5	–

Table 10.1 (continued)

Country	Area (000 sq km)	Population (millions)	International tourists (thousands)
Togo	56.8	6.4	–
Tunisia	163.6	10.1	6,550
Uganda	241	29.9	539
Western Sahara	–	–	–
Zambia	752.6	11.7	–
Zimbabwe	390.8	13.2	2.287

Source: UNWTO (2007); World Bank (2007a, 2007b).

interface of an array of resources from four historical periods: the pre-historical elements based on Africa as the cradle of humankind; relics of traditional African kingdoms and civilizations; exogenous cultural elements derived from Arab adventurism and European colonization; and elements since the late 1950s post-colonial period. As a result of geographical and historical influences, Africa hosts 116 of the 878 World Heritage properties in the world (see Figure 10.1). SubSaharan countries with significant numbers of properties include: Ethiopia (8), South Africa (8), Tanzania (7), Senegal (5), and Zimbabwe (5).

The geographical, historical, and cultural diversity of Africa makes it a region with enormous potential for economic development using tourism as a tool for diversification beyond the principal traditional economic activities. The region remains arguably the poorest on all economic and human development indices. For example, both the UNDP (2008) and the World Bank (2007b) rank subSaharan Africa as the world's poorest region with respect to per capita gross national product (GNP), low literacy rates, high infant mortality, and relatively short life expectancy. The region also has some of the fastest population growth rates in the world with populations doubling every twenty to twenty-five years. Political factors such as frequent and protracted civil wars, military interventions (Teye 1988), and corruption have combined with devastating health and medical issues (such as HIV/AIDS) and natural disasters (drought and famine) to curtail economic development. Yet, there is widespread belief that tourism can contribute in significant ways to sustainable economic development and, indeed, can help to alleviate poverty in most, if not all, African countries. The Sustainable Tourism–Eliminating Poverty (ST–EP) Program of the UNWTO is aimed at the least developed countries, especially those in Africa. Its main objective is "to contribute to poverty reduction through the establishment of community-based tourism development projects that respect the environment and benefit the most disadvantaged population" (WTO 2005).

Over the last twenty-seven years, tourism in Africa as a whole has been described by the UNWTO as being "quite positive" (WTO 2005). The number of international arrivals more than doubled between 1980 and 1990

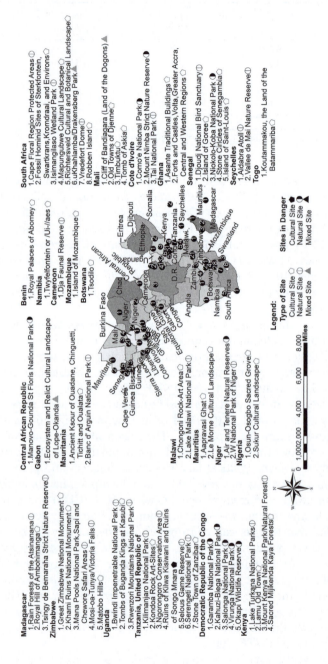

Madagascar
1. Rain Forests of the Atsinanana⊕
2. Royal Hill of Ambohimanga◯
3. Tsingy de Bemaraha Strict Nature Reserve◯

Zimbabwe
1. Great Zimbabwe National Monument◯
2. Khami Ruins National Monument◯
3. Mana Pools National Park,Sapi and Chewore Safari Areas◯
4. Mosi-oa-Tunya/Victoria Falls◯
5. Matobo Hills◯

Uganda
1. Bwindi Impenetrable National Park⊕
2. Tombs of Buganda Kings at Kasubi⊕
3. Rwenzori Mountains National Park◯

Tanzania, United Republic of
1. Kilimanjaro National Park◯
2. Kondoa Rock Art-Sites◯
3. Ngorongoro Conservation Area◯
4. Ruins of Kilwa Kisiwani and Ruins of Songo Mnara ●
5. Selous Game Reserve◯
6. Serengeti National Park◯
7. Stone Town of Zanzibar◯

Democratic Republic of the Congo
1. Garamba National Park●
2. Kahuzi-Biega National Park●
3. Salonga National Park●
4. Virunga National Park●
5. Okapi Wildlife Reserve●

Kenya
1. Lake Turkana National Parks⊕
2. Lamu Old Town◯
3. Mount Kenya National Park/Natural Forest◯
4. Sacred Mijikenda Kaya Forests◯

Central African Republic
1. Manovo-Gounda St Floris National Park●

Gabon
1. Ecosystem and Relict Cultural Landscape of Lope-Okanda▲

Mauritania
1. Ancient Ksour of Ouadane, Chinguetti, Tichitt and Oualata◯
2. Banc d' Arguin National Park◯

Malawi
1. Chongoni Rock-Art Area◯
2. Lake Malawi National Park◯

Mauritius
1. Aapravasi Ghat◯
2. Le Morne Cultural Landscape◯

Niger
1. Air and Tenere Natural Reserves●
2. W National Park of Niger◯

Nigeria
1. Osun-Osogbo Sacred Grove◯
2. Sukur Cultural Landscape◯

Benin
1. Royal Palaces of Abomey ◯

Namibia
1. Twyfelfontein or /Ui-//aes ◯

Cameroon
1. Dja Faunal Reserve◯

Mozambique
1. Island of Mozambique◯

Botswana
1. Tsodilo◯

South Africa
1. Cape Floral Region Protected Areas ⊕
2. Fossil Hominid Sites of Sterkfontein, Swartkrans,Kromdraai, and Environs◯
3. Isimangliaso Wetland Park ◯
4. Mapungubwe Cultural Landscape◯
5. Richtersveld Cultural and Botanical Landscape◯
6. uKhahlamba/Drakensberg Park▲
7. Vredefort Dome◯
8. Robben Island◯

Mali
1. Cliff of Bandiagara (Land of the Dogons)▲
2. Old Towns of Djenne◯
3. Timbuktu◯
4. Tomb of Askia◯

Cote d'Ivoire
1. Como'e National Park●
2. Mount Nimba Strict Nature Reserve●
3. Tai National Park ⊕

Ghana
1. Asante Traditional Buildings◯
2. Forts and Castles,Volta,Greater Accra, Central and Western Regions ◯

Senegal
1. Djoudj National Bird Sanctuary◯
2. Island of Goree◯
3. Niokolo-Koba National Park●
4. Stone Circles of Senegambia◯
5. Island of Saint-Louis ◯

Seychelles
1. Aldabra Atoll◯
2. Vallee de Mai Nature Reserve◯

Togo
1. Koutammakou, the Land of the Batammariba◯

Legend:

Type of Site
Cultural Site ◯
Natural Site ◯
Mixed Site ▲

Sites in Danger
Cultural Site ●
Natural Site ●
Mixed Site ▲

Figure 10.1 Developing countries and World Heritage Sites in subSaharan Africa

from only 7.3 million to 15 million. The arrival numbers doubled again in the following decade between 1990 and 2000, and actually grew at an average annual rate of 6.4 percent compared with the world average of 4.2 percent (WTO 2000). In recent years, Africa's tourism growth rate continues to outpace those of the rest of the world. For example, the continent recorded 9 percent growth in 2006 compared with 5.4 percent for the world (UNWTO 2007). In spite of these encouraging overall growth rates, Africa's share of the 846 million international tourist arrivals in 2006 was only 40.7 million. Similarly, international tourist receipts in the same year were only US$24.3 billion out of the total of US$733 billion. These figures represent 4.8 percent and 3.3 percent, respectively, of the global international tourism industry. Further examination also indicates that the industry is concentrated in relatively few countries. South Africa (20.6 percent), Morocco (16.1 percent), and Tunisia (16.1 percent) collectively receive more than 50 percent of all arrivals in Africa. Other important countries in subSaharan Africa are Kenya, Tanzania, Namibia, and Botswana, but it is important to note that, even though it is covered in Chapter 9, Egypt received 8.65 million arrivals in 2006. This figure is more than South Africa's 8.39 million, which actually makes Egypt the leading tourism destination on the African continent. It also means that Africa's tourism figures are underestimated and were actually 49.35 million arrivals and US$31.89 billion in 2006 (UNWTO 2007).

Against this background, this chapter examines contemporary heritage tourism in Africa and, because of space limitations, it focuses primarily on the region's built heritage. Given that Chapter 9 is devoted to North Africa (and Southwest Asia), this chapter's primary focus is subSaharan Africa. Nonetheless, as the geographical, historical, and cultural influences of the two regions are closely interwoven, it is impossible to completely exclude a discussion on North Africa. For example, geography and history came together along the River Nile from Ethiopia's Abyssinian Highlands to evolve the Egyptian civilization and those that emerged later as Kush, Axum, and Meroe in today's Sudan and Ethiopia (Knight and Newman 1976). Their relics, monuments, and heritage are important to subSaharan Africa. Similarly, Islam spread south from North Africa and linked with commerce (the transSahara salt and gold trade) providing diverse religious heritage that affects nearly half of subSaharan Africa's population. The approach used here is to utilize Africa's triple (tripartite) heritage (Mazrui 1986), consisting of traditional or indigenous African heritage, Islamic heritage, and European colonial heritage in examining heritage tourism in Africa. Heritage as used in this chapter refers to both tangible (physical) and intangible items of value from a society's past creation and inheritance that have been passed on to the present generation. Key cultural resources that may be incorporated into heritage tourism have been classified elsewhere (Prentice 1993), but may include historical sites or buildings (forts and castles, religious shrines), or cultural heritage (festivals, arts and crafts). If properly developed, managed, presented, and interpreted, heritage tourism in Africa can offer great economic and cultural benefits for host destinations, in addition to visitors deriving cultural, personal, emotional, spiritual, and educational benefits.

Overview of heritage tourism in Africa

Early forms of tourism in Africa can be traced far back to the Roman occupation of Egypt beginning around 30 BCE (Appiah and Gates 1999). The Romans explored the ruins of Thebes and tombs in the Valley of the Kings. They were followed by Arab, Asian, and later European explorers such as Livingstone, Burton, and Speke (Garfield 1994; Huggins and Jackson 1969). Followers of Islam and Christian missionaries also traveled extensively in subSaharan Africa. The colonial period laid the foundations for nature-based tourism in the forms of various safaris in East and Southern Africa (Ouma 1970). However, cultural tourism is a very significant component of contemporary tourism in Africa, in spite of the industry's relatively small size compared with other regions, as well as the size of the continent.

Egypt, which boasts some of the continent's most ancient and magnificent heritage attractions, including monuments and artifacts, has historically received cultural and heritage tourists, most of whom come to see the Valley of the Kings, Thebes, the Nile, and the museums in Cairo. Other North African locations such as Casablanca, Marrakech, Fèz, Agadir, and Carthage have established North Africa as the continent's leading tourism subregion. Obviously, a critical factor is North Africa's proximity to Europe which translates into shorter travel time, lower cost, as well as choice of waterfront locations on the Atlantic coast, Mediterranean Sea, and the Red Sea. However, as the discussion will show later, there are significant historical, cultural, and religious connections between Europe, North Africa, and subSaharan Africa that have resulted in important common heritage across the whole continent.

Since the late 1970s, heritage tourism has expanded to encompass a broader range of activities in subSaharan Africa. Africans of the diaspora around the world have become a key market segment for heritage tourism in Africa. From South America, the Caribbean, Central America, the United States, and Europe, especially the United Kingdom and France, today's descendants of African slaves have become interested in their black heritage including such cultural manifestations as language, music, dance, arts, craft, movies, books, and cuisine. For those who undertake the journey back to the motherland or homeland, they seek to make that connection as pilgrims to spiritually, emotionally, and psychologically "find" themselves and search for where it all began for them centuries and generations ago (Austin 2000; Timothy and Teye 2004). For most Americans, Alex Haley's (1976) book, *Roots*, and the subsequent mini-TV series stirred up that innermost yearning or the motivational pull to travel to the African continent. Because *Roots* was based on historical events traced to Jufurre, a small village in the Gambia, West African countries in particular have benefited from the flow of heritage tourists of African descent. This, in turn, has led to a number of important heritage tourism development projects in that subregion. Ghana, Benin, and Gorée Island, located off the coast of Senegal, were all major trans-Atlantic slave ports. Ghana's tourism industry has benefited the most from heritage tourism development in West Africa. Three of its former slave forts were declared World Heritage Sites: Cape Coast Castle, Elmina Castle, and Fort Saint Jago. The core project for the country's 15-year Tourism Development Plan combined heritage and nature-based tourism in two communities (Cape Coast and Elmina) where the three slave forts are located. Referred to as the Natural Resource and Historic Restoration Project, it was funded by the United Nations Development Program (UNDP) and the United States Agency for International Development (USAID). Today, Cape Coast, Elmina, and the recently developed forest animal park at Kukum are the leading tourism destinations in Ghana (Teye 2008; Teye *et al.* 2002). There is also an annual Pan-African and Emancipation Festival (PANAFEST) in Cape Coast that attracts large numbers of tourists from the African diaspora. In the Gambia, the village of Jufurre has become a pilgrimage site for many African American heritage tourists. The Gambia holds an annual Roots Homecoming Festival, which celebrates the heritage links between diasporic Africans.

Heritage tourism has also become increasingly important in the Southern and East Africa. Visitors to South Africa, for example, can spend a night in a traditional home of a Xhosa, Sotho, Pedi, or Zulu family in the Lesedi Cultural Village outside Johannesburg. Residents of the village wear the traditional dress reflecting Xhosa or Zulu heritage. Story telling, musical and dance performances also serve to educate visitors while preserving their heritage and also passing this on to the next generation. Similar cultural attributes have been incorporated into the tourism industry in Botswana, Lesotho, Swaziland, Mozambique, Malawi, Zimbabwe, and Zambia. The ruins of the old Zimbabwe Civilization have become a major tourism destination in Southern Africa that are being preserved with revenue generated by visitors.

Indigenous African heritage

Ehert (2002: 3) succinctly describes Africa's unique position and role in the history, culture, and heritage of all humanity:

> Africa lies at the heart of human history. It is the continent from which the distant ancestors of every one of us, no matter who we are today, originally came. Its peoples participated integrally in the great transformations of world history, from the first rise of agricultural ways of life to various inventions of metal-working to the growth and spread of global networks of commerce ... the African continent presents us with a historical panorama of surpassing richness and diversity.

Archeological discoveries in various parts of Africa over the past seventy years have provided scientific evidence that human genesis or origins can be traced to the continent. The first major evidence came in 1931 when Louis Leakey discovered *Proconsul*, an ape-like creature, in the Lake Victoria region of Kenya. Almost forty years later in 1959, Louis and his wife Mary Leakey discovered *Zinjanthropus* in the now world-famous archeological site at Olduvai Gorge in Tanzania (Mazrui 1986). Since then, several important discoveries have provided evidence suggesting that the natural and evolutionary heritage can be traced to Africa. Among the significant archeological discoveries are:

1) The discovery of *Homo habilis* by Louis Leakey in 1960 in Olduvai Gorge in northern Tanzania;
2) The *Man from Kibish* discovered in 1967 on the shores of the River Kibish in Ethiopia; and
3) *Homo erectus* discovered in 1984 on the shores of Lake Turkana in Ethiopia by Kamoya Kimeu and Richard Leakey (the son of Louis and Mary Leakey).

Scientific work to trace human evolution and migration to other continents has also helped to establish Africa as a major destination for archeology-based tourism, not only in East Africa but across the continent. For example, excavations in West and North Africa explore the trans-Sahara salt and gold trade, while others examine old civilizations in Northern, West, East, Central, and Southern Africa. Table 10.2 summarizes the main African kingdoms and civilizations that existed prior to European incursions or the "discovery" of Africa. These kingdoms, empires, and civilizations are important for a number of reasons. First, they symbolize pure and unadulterated indigenous African heritage that is critical to the identity of Africans. Second, many of these civilizations produced structures and relics that have global significance, resulting in them being placed on the World Heritage List. These range from major tourist attractions such as the pyramids and temples of the Egyptian

Table 10.2 Major indigenous African civilizations and empires

Empire	Location	Period
Nubian	Nile Region (Egypt/Sudan)	5000–3000 BC
Egyptian	Nile Region	3000–2000 BC
Kushite, Meroe	Egypt/Sudan (North Africa)	500 BC–320 AD
Axum	Sudan (North Africa)	230 AD
Ghana	Central West Africa	800 AD
Zimbabwe	Southern Africa	850 AD
Mali	Central West Africa	1240 AD
Songhai	West Africa	1335–1495 AD

Sources: Ehert (2002); Huggins and Jackson (1969); Mazrui (1986); Appiah and Gates (1999).

civilization to structures less known in the West, such as the Asante Traditional Buildings of the "great Ashanti civilization that reached its high point in the 18th century" (UNESCO 2008b). Third, these civilizations and their rich natural resources provided some of the reasons that attracted Christian missionaries, Islamic traders, European explorers, and colonial powers to Africa. The activities of each group, singularly and in combination, have altered the historical and cultural landscape of the continent as well as its people, thereby influencing contemporary and future heritage tourism development. Ultimately, a discourse on heritage tourism in Africa is a complex exercise focusing on the continent's global linkages through which diverse cultural influences manifest themselves to Africans, diasporic Africans, Europeans, Arabs, and Asians. As will become obvious later in this discussion, Asian heritage in Africa can be seen particularly in countries such as South Africa, Kenya, and Uganda. Even more dominant Asian influences in heritage tourism can be seen on the island destinations of the Indian Ocean including Madagascar, Seychelles, Mauritius, and Comoros.

The indigenous African civilizations of Nubia, Egypt, Kush, Meroe, and Axum in parts of areas currently occupied by Egypt, Sudan, and Ethiopia all had significant cultural and technological impacts on Africa, and indeed, the rest of the world. These include architecture, agriculture, mathematics, astronomy, astrology, language, writing, religion, traditions, music, dance, and social norms. A brief review of the World Heritage List noted in Figure 10.1 indicates the wealth of resources and attractions handed down by these civilizations. A brief description of a few of these civilizations will serve to provide a better understanding of their contributions to heritage tourism in subSaharan Africa.

The ancient African civilization of Egypt was centered in the Nile Delta and the lower Nile Valley. It is regarded as one of the pivotal foundations of Western culture, but also steeped in both Middle East and African cultures. This civilization covered a very long period and was divided into the Old, Middle, and New Kingdoms with intermediate periods, followed by the late and Ptolemaic periods up to the conquest of Alexander the Great. The result of this long period of sustained civilization is the endowment of the diversity and extraordinarily large numbers of unique cultural heritage, which partly explains Egypt's position as the leading tourism destination in Africa. Among the key heritage tourism destinations is the Valley of the Kings. During the Old and Middle Kingdoms between 2980 and 1580 BCE, the pharaohs commissioned pyramid tombs and temples in anticipation of their journeys to the afterlife. The tombs were filled with the valuables considered necessary for life after death, including jewels, precious metals, food, tools, furniture, and even royal servants and pets (Appiah and Gates 1999). Archeological evidence indicates that work on a pharaoh's tomb began the day he ascended the throne and ended the day he died. The result is the existence of some of the most complex and elaborate tombs anywhere in the world. For example, Amenhotep I had his temple and tomb built into the side of the limestone cliffs in the valley, with deep corridors stretching as much as 100 meters below the earth. More than sixty such tombs have been rediscovered since the eighteenth century. Perhaps the most extraordinary discovery was the tomb of the boy king Tutankhamen in 1922. Although it is by far the smallest tomb in the valley, archeologists recovered more than 5,000 artifacts from the tomb, many of which now reside in the Cairo Museum.

The Valley of the Kings is among Egypt's greatest heritage tourist attractions, and has been for millennia, as evidence from ancient Greek and Roman visitors testify. The continuing fascination with ancient Egypt attracts an estimated 3,000 visitors per day to the valley sites. It is important to note that the Valley of the Kings is not one of the seven World Heritage Sites in Egypt. That single site on the list includes other significant and popular attractions such as Luxor and the Valley of the Queens. The World Heritage location is classified as Ancient Thebes with its Necropolis, which is officially described as "Thebes, the city of the god Amon, was the capital of Egypt during the period of the Middle and New Kingdoms. With the temples and palaces at Karnak and Luxor, and the necropolises of the Valley of the Kings and the Valley of the Queens, Thebes is a striking testimony to Egyptian civilization at its height" (UNESCO 2008c).

Colonial heritage

While the colonial period effectively began after the partitioning of Africa at the Berlin Conference of 1884, there had been a long period of contact between Africa and Europe. For example, the Greeks and Romans traded with ancient Egypt, which later led to a flourishing trade in the Mediterranean. Today, Greco-Roman influence and heritage can be seen in parts of North Africa such Carthage in modern-day Tunisia. Active European presence in Africa began in earnest after the Renaissance and the Age of Discovery. Table 10.3 shows some of the European explorers and early international tourists who "discovered" geographical landmarks such as rivers, lakes, waterfalls, and mountains. They named some of these discoveries after their kings, queens,

Table 10.3 Early European explorers in Africa

Explorer	Country of origin	Exploration period
Vasco da Gama	Portugal	1497–99
Bartholomew Diaz	Portugal	1481–87
Mungo Park	Scotland	1795–97
Hugh Clapperton	Scotland	1822–27
David Livingstone	Scotland	1841–73
John Speke	England	1857–63
James Grant	Scotland	1860–63
Morton Stanley	England	1871–89

Source: Compiled from Appiah and Gates (1999).

princes, and princesses. Noteworthy among these are the Victoria Falls, a major tourist attraction on the Zambezi River, Lake Edward, and Lake Rudolf. It is noteworthy that the initial phase of formal international tourism in Africa involved predominantly European adventurers attracted by "wild" Africa, who were eager to experience these discoveries during the early phase of the establishment of European heritage on the continent (Mazrui 1986; Ouma 1970).

The accounts of these explorers, which detailed the natural resource wealth for Europe's expanding industries, eventually led to the late nineteenth-century "scramble for Africa" (Ehert 2002; Mazrui 1986). The rush to have a presence in Africa was so intense that European leaders with little prior interest in colonization, such as German Chancellor Otto von Bismarck, staked claims to the continent. At one point, King Leopold of Belgium claimed all the vast territory (2.43 million square kilometers) of today's Democratic Republic of Congo as his private game preserve for hunting. At the Berlin Conference of 1884–85, European leaders agreed to partition Africa, thereby bringing almost 90 percent of Africa under European control. Four countries (France, Great Britain, Portugal, and Belgium) effectively became colonial powers administering their territories until the late 1950s (Curtin *et al.* 1995). Their influences and heritage are entrenched and widespread through imposed "official" languages, architecture, and various systems of education, administration, forms of government, laws, place names, and religion, among others. Owing to the intense military, political, religious, as well as commercial and human trading activities, some European countries that did not participate in the colonial period still have important heritage across subSaharan Africa. For example, German heritage is still prevalent in Namibia and parts of Ghana and Togo. Tanzania was actually known during the colonial period as "German East Africa," even long after it had become a British territory. Spanish, Dutch, and Danish heritage still flourishes in parts of West Africa, especially in communities along the coast. For instance, the Dutch and Danish built several castles and trading posts in Ghana where Dutch names such as Van Dyke are still common in many households.

There are a number of important outcomes of European influences relevant to heritage tourism that warrant mentioning briefly. First, some Europeans were attracted to the highlands of East Africa and Southern Africa for geographical reasons. For example, a large number of British citizens settled in Kenya, Uganda, Tanzania, and Zimbabwe. Their presence has led to a highly embedded British heritage and attractions in these areas compared with, for example, West Africa, where high temperatures and humidity cause greater incidence of tropical diseases, such as malaria. The region became known as the "white man's grave" and was devoid of any major European settler populations. European heritage in East and Southern Africa is in the form of large coffee and tea plantations, which are important today for agro-tourism, large numbers of game parks for eco-tourism, and even the countries of Northern and Southern Rhodesia that were renamed Zambia and Zimbabwe, respectively, after each became politically independent. These countries were originally named after Cecil Rhodes, the British industrialist who is buried in today's Zimbabwe. Today, large numbers of descendants of the original white immigrants known as the "pioneer column" settlers (that Rhodes led) travel on pilgrimage to Zimbabwe. Second, the British "imported" Asians from their colonies in the Indian subcontinent, and their contemporary influence and heritage are widespread in Kenya, Uganda, Tanzania, and South Africa. These influences are seen in their language, religion, temples, music, dance, and commercial activities. Third, Dutch settlers in South Africa provided a unique case of the Dutch and British in a racially turbulent and mostly confrontational relationship with African cultures. Today, African heritage (including Bantu, Zulu, Shona) is being preserved alongside European, Asian, and interracially mixed cultures. Indeed, places such as Robben Island in South Africa, where Nelson Mandela was imprisoned for nearly thirty years for his opposition to apartheid, are heritage tourism destinations for both domestic and international tourists who visit the popular museum on the island. Located only 12 kilometers from Cape Town, its notoriety as a dark heritage attraction lies in its 400-year history as a place where opponents of European supremacy were banished for life. Finally, a large number of Africans emigrated to European countries, especially to the United Kingdom, France, and Belgium. They have become part of the African diaspora, constituting a significant market for Africa's heritage tourism (Harris 1982). This market segment deserves further examination later in this chapter.

Religious influences in heritage tourism

This section discusses heritage dimensions of tourism from the influences of the traditional African, Arabian Islam, and European Christianity. Religion is central to African cultures right from birth, throughout life, to death and beyond. Traditional African religions are as diverse as the continent's tribes and clans. Despite centuries of contacts with other dominant religions, such as Christianity and Islam, which have far greater resources, appeal, and significant external financial support, Africans still have hundreds of distinct religions. African religious thought is expressed through the recitation of myth and oral traditions, and through discussion both among elders and between generations. In this respect, religion plays a vital role in domestic tourism in Africa, largely with respect to visiting friends and relations, funerals, and visits to religious shrines. Religion is also expressed through rituals, which often involves making offerings to attract a spirit's power or win its benevolence (Appiah and Gates 1999). Traditional African religions have significant relevance for heritage tourism development in Africa. First, many African art forms that are purchased by tourists are used primarily in religious rituals by locals. In most of Africa, masks and costumes are used to impersonate the lesser spirits. For example, among the Yoruba and Igbo of

Nigeria and the Ewe of Togo and Ghana, wearing a mask and costume invites the presence of a god into one's body. Second, through the slave trade, traditional African religions now have large followings outside the continent in South America, the Caribbean, and Central America. For example, Voodoo is practiced in Haiti and some parts of the United States (Louisiana). It is common to find followers of such traditional religious practices traveling the heritage trail to shrines in Benin, Togo, and Ghana. Third, some of these shrines are heritage sites on the UNESCO List. A good example is the Cliffs of Bandiagara (also known as the Land of the Dogons) in Mali that has religious attributes with respect to ancestral worship (Shackley 1999).

> The Bandiagara site is an outstanding landscape of cliffs and sandy plateaux with some beautiful architecture (houses, granaries, altars, sanctuaries and *Togu Na*, or communal meeting-places). Several age-old social traditions live on in the region (masks, feasts, rituals, and ceremonies involving ancestor worship). The geological, archaeological and ethnological interest, together with the landscape, make the Bandiagara plateau one of West Africa's most impressive sites.
>
> UNESCO (2008c)

The influence of Islam on tourism is profound (Timothy and Iverson 2006) and in Africa is deeply rooted. It actually predates the European period and Christianity. Cultural influences from the Arabian Peninsula of the Middle East can be traced back once again to the period of the ancient Egyptian civilization. About 40 percent of Africa's total population of about 900 million are Muslims. The initial influence was in North Africa, then East Africa in what became known as the Swahili Coast. Today, much of West Africa, especially countries such as Mali, Chad, Burkina Faso, Niger, and Mauritania, are principally Islamic. Given its long history and broad geographical reach on the continent, Islam has left significant cultural heritage including architecture (mosques), education, laws (Sharia), festivals (Ramadan), and religion (five daily prayers). Islam was embraced in the desert regions of the Sahara and played a central role in establishing centers of learning as well as the trans-Saharan gold and salt trade during the periods of the Mali and Songhai Empires. For example, Timbuktu in Mali was the home of the prestigious Koranic Sankore University and other *madrasas*. It was also an intellectual and spiritual capital and a center for the propagation of Islam throughout Africa in the fifteenth and sixteenth centuries. Its three great mosques, Djingareyber, Sankore, and Sidi Yahia, recall Timbuktu's golden age (UNESCO 2008d).

Another area in Mali illustrates the influences of Islam in the country. The dramatic 17-meter pyramidal structure of the Tomb of Askia was built by Askia Mohamed, the Emperor of Songhai, in 1495 in his capital Gao. It bears testimony to the power and riches of the empire that flourished in the fifteenth and sixteenth centuries through its control of the trans-Saharan trade, notably in salt and gold.

It is also a fine example of the monumental mud-building traditions of the West African Sahel. The complex, including the pyramidal tomb, two flat-roofed mosque buildings, the mosque cemetery and the open-air assembly ground, was built when Gao became the capital of the Songhai Empire and after Askia Mohamed had returned from Mecca and made Islam the official religion of the empire.

UNESCO (2008e)

One significant aspect of the influence of Islam on tourism is the large annual *Hajj* pilgrimage to Mecca (Ahmed 1992), in which more than one million African Muslims participate as part of the worldwide tourism phenomenon deeply rooted in religious heritage that links Africa to the Arabian Peninsula and the rest of the Islamic world.

The role of religion in tourism, and in particular Christianity, within heritage tourism has been well documented (Timothy and Olsen 2006) but to a lesser extent in Africa. Christianity pervades the lives of Africans, with leading international clerics from Africa in the leading churches, such as the Roman Catholic, Anglican, and Protestant churches. Early Christianity on the African continent has been traced to missionaries and traders to subSaharan African countries. From the fifteenth century onward, Europeans intensified their exploration, trading, and colonizing activities in Africa. As an evangelizing religion, Christianity's missionaries were important in "spreading the gospel." European governments and trading companies often supported missionary work in Africa. The Portuguese example illustrates the intimate relationship between the church and state (crown) and business (commerce). In the Treaty of Tordesillas (1494), the Pope recognized Portuguese claims to Africa. The crown was also responsible for attempting to convert the indigenous people to Christianity. Much of the missionary effort over the next two and a half centuries was conducted under Portuguese authority. Trade was also linked to religion, as evidenced by the sentiments of David Livingstone, the Scottish missionary who traveled widely in Southern and Central Africa in the mid-nineteenth century, declaring in 1857 that Africa needed Christianity and commerce (Appiah and Gates 1999).

In addition to evangelizing, mission societies often provided social services and resources for Africans, such as schools and hospitals. One religious edifice considered a major religious heritage tourist attraction is Our Lady of Peace of Yamoussoukro Basilica in the Ivory Coast. Built by the country's former head of state, Felix Houphouet-Boigny, in his home town, which became the country's capital, it is estimated to have cost between US$150 and US$800 million. It is the world's largest church, surpassing even St. Peter's Basilica in Rome, and was once visited by the Pope.

Besides the three major religions, there are others that have significance for religious heritage tourism. These range from Hinduism and Buddhism to Judaism and Rastafarianism whose followers live mostly in the Caribbean but believe the former Emperor Haile Salassie was their God or (Jah). The emperor, whose real name was Ras Fatari, was also known as the King of Judah and believed to be a direct descendant of King Solomon.

Chevannes (1994)

The black diaspora and heritage tourism

So far, the discussion has focused on the supply side of heritage tourism, or what constitutes the stock of attractions that are likely to motivate both visitors and residents to participate in heritage tourism in Africa. However, it is important to examine the demand side of heritage tourism, in particular the slave trade heritage, which links traditional Africa with Europe, the Americas and the Middle East. Arguably, Africa's largest market for heritage tourism is the large and diverse market segments of the black diaspora, especially its members in the United States (Essah 2001; Goodrich 1985). It is an established fact that there were two distinct regional slave trading traffic routes from Africa for several centuries prior to the first significant European contacts with Africa in the fourteenth century. For example, the trans-Sahara slave trade was from Western and Central Africa to North Africa, while the Red Sea slave trade exported Africans from Eastern Africa to Arabia and South Asia. Africa's contact with Europe established the third wave of slave traffic, consisting of the trans-Atlantic slave trade from West Africa to the Americas and Europe. This was by far the most dominant region in terms of the number of slaves captured, trafficked, and sold, the duration of the slave raids, trading and slavery, as well as the impact on both sides of the Atlantic Ocean. Figures are hard to establish, but some scholars suggest that, while the trans-Sahara and Red Sea trade involved about 6 million slaves, in the trans-Atlantic slave trade "from the 1520s to the 1860s, an estimated 11 to 12 million African men, women and children were forcibly herded onto European vessels for a life of slavery in the Western Hemisphere" (Appiah and Gates 1999: np). Over a period of about 350 years, an estimated 9 to 10 million African slaves survived the trans-Atlantic voyage as human cargo to be purchased by white plantation owners and slave traders in the New World of the Caribbean, Latin America, and North America.

While African slaves were sold in slave markets as far north as New England and as far south as present-day Argentina, today's descendants of African slaves are spread all over the world. It is this global dispersal of peoples of African heritage that has come to be known as the black diaspora. With the exception of a few freed slaves who were returned to Sierra Leone and Liberia by England and the United States, respectively, following the abolition of slavery, the vast majority of today's slave descendants are concentrated in the Caribbean and North, Central, and South America, as well as in a few European countries such as the United Kingdom and France (Harris 1982).

African Americans constitute a tremendous demand for Africa's heritage tourism supply for a number of reasons (Timothy and Teye 2004). First, with a population of about 40 million, they are the single largest group of African descent and heritage in any country. Second, their average income level has experienced annual growth rates of about 16 percent between 1990 and 2000, currently representing a US$450 billion market. As a result, the African American market was considered the fourteenth largest market about ten years ago with more disposable and discretionary income than Australia, Mexico, or Russia (Malveaux 1998). Third, African Americans predominantly constitute a single linguistic market. Unlike peoples of African descent and heritage in many countries in the Caribbean and Latin America who speak various languages, including English, French, Spanish, Portuguese, and various derivations of local and European languages (such Patois, Creole, and Papiamento), the primary language of African Americans is English. This factor tends to facilitate and enhance their travel experiences in Africa, including travel to the large number of Francophone countries in West Africa. All the major French-speaking countries in subSaharan Africa share, at least, one common border with an English-speaking country. Fourth, the growth and significance of the African American market for both domestic and international travel has been recognized by key sectors of the travel industry including hotels, airlines, cruise lines, and theme parks. Finally, the economically prosperous African American population tends to be concentrated in large metropolitan areas, which are also gateways for international travel.

The large number of historical structures built in connection with the slave trade constitute a critical component of a relatively new but major heritage tourism project of international significance initiated in the early 1990s. It is envisaged to have tremendous tourism potential for Africa and the rest of the world. At a joint meeting of the World Tourism Organization (UNWTO) and the United Nations Educational, Scientific and Cultural Organization (UNESCO) held in Accra, Ghana, in April 1995, the decision was formalized "to rehabilitate, restore and promote the tangible and intangible heritage handed down by the slave trade for the purpose of cultural tourism, thereby throwing into relief the common nature of the slave trade in terms of Africa, Europe, the Americas and the Caribbean" (WTO/UNESCO 1995). This resolution was formally adopted at the 27th meeting of the UNWTO Regional Commission for Africa in 1995 in Durban, South Africa. This program, which has become known as the Slave Route Project, is directly linked to heritage tourism development, within the context of the preservation and restoration of world sites and corridors:

> In the final analysis, the program aims to forge a close link between the ethical exigencies of preserving the memory of the slave trade, which historians now consider "the biggest single tragedy in the history of man on account of its scope". ... The forts and castles on the coast of Ghana, in particular, Cape Coast, Elmina and l'Ile de Gorée on the Senegalese coast, symbolize these memorial sites. And now that they have been placed on the list of UNESCO's world heritage, their preservation,

restoration and promotion henceforth forms part of the universal heritage of mankind. It is also the case of other sites like Ouidah (Benin), Angola, Mozambique, Tanzania, etc.

WTO/UNESCO (1995)

The growing interest in heritage tourism in Africa is reflected in the Cairo Declaration of 1995, which was adopted under the auspices of UNWTO and UNESCO. A principal objective was for African countries to identify, develop, and preserve a number of World Heritage Sites in cooperation with international agencies and special interest groups. Ultimately, it was expected that such developments would enhance the promotion of cultural resources and attractions, which would, in turn, lead to the sustainable development of heritage and ethnic tourism as part of the continent's overall economic development strategy.

Conclusions

Heritage tourism development through the identification and restoration of cultural resources can have tremendous economic, cultural, and environmental benefits for African countries. Heritage tourism development has helped save large numbers of heritage sites from floods in the Nile Valley and from desertification in the Sahel region of Africa. Several issues can be identified as African countries integrate heritage tourism into their overall tourism development strategies. First, a significant number of unique and valuable artifacts from Africa that were taken away by colonialists still reside in private hands and museums in Europe. A good example is the Aksum Obelisk, which was taken to Rome in 1937 by Mussolini's troops. Given the significance of the 152-ton monolith, which was sculpted seventeen centuries ago, UNESCO negotiated and funded its return to Ethiopia in 2008. Not all countries have the resources of leverage to negotiate such an undertaking. For example, to accomplish the return of the Obelisk, the world's largest aircraft, the six-engine Antonov, was chartered, and the Aksum airport had to be modernized to accommodate the aircraft and delivery (UNESCO 2008f).

The second issue deals with the question of whose heritage it is and how to develop it (Tunbridge 1997). It is clear that Africa's triple heritage derived from traditional Africa, the Middle East, and colonial Europe provides opportunities as well as challenges. There are several important policy issues, as well as those that relate to the politics of heritage tourism development. These can be at the local community, national, and international levels. For example, what are the expectations of African Americans when visiting Africa? What are their experiences when they visit the slave forts and other heritage sites? How do these experiences compare with those of white visitors? What are the group dynamics between white and black tour groups visiting these heritage sites at the same time or as a single but diverse racial group? How do African Americans perceive their African hosts? Other issues that relate to visitors' activities such as taking photographs of locals, residents' lack of awareness and involvement in the development of their cultural heritage, and black visitors' anger at the "whitewashing" of their history and heritage through restoration activities. An example from Ghana illustrates the issue:

> For many African-Americans, the castles are sacred grounds not to be desecrated. They do not want them to be made beautiful or to be white-washed. They want the original stench to remain in the dungeons ... a return to the slave forts for Diaspora Blacks is a "necessary act of self-realization" for the "spirits of the Diaspora are somehow tied to these historic structures." Some Diaspora Blacks feel that even though they are not Ghanaians, the castle belongs to them.
>
> Bruner (1996: 291)

The third issue relates to the authenticity of the presentation and interpretation of the historical events associated with these heritage sites and structures. The restoration projects were carried out principally by foreign consultants and international agencies such UNESCO, Conservation International (CI), and the Smithsonian Institution. Austin (1997, 2000) examined some of the critical issues in the development of historical and heritage structures with specific reference to Ghana's Elmina Castle and concluded that, for all of Africa:

> How ordinary Africans as the destination hosts, see and relate to the African Diaspora in particular and how project organizers view those undertaking the visit, whether as "tourists," "visitors," "pilgrims," or even "foreigners" is of major importance. Africans in the Diaspora on visits to the African continent, see themselves as "coming home." This feeling of "coming home" and the reconnection with the land of their fathers represents the essence of their visit. In reality, most are received as strangers and treated as such by the host.
>
> Austin (2000: 213)

Like the Silk Route Project in Asia, the Slave Route Project is multinational in scope, but it is on a much larger scale and, above all, has necessary emotional significance for the market segment that African countries are attempting to attract as heritage visitors. This is a new development area for African countries, and clear policies need to be established for a balanced development, presentation, and interpretation of the cultural resources, as well as the sensitization and education of developers, tourism operators, residents, and visitors, both black and white. Careful planning, preparations, and training must be involved. As Austin (1997: 214) states:

> ... as a result of the sensitive nature of the events of the [slave] trade to various groups, intergroup conflicts are inevitable at sites and other presentations associated with it. These conflicts over time may shape the

future market and the viability of the tourism development. Management ... must seek to identify, understand, and manage these conflicts.

For example, a study by Teye and Timothy (2004) found that local tour guides in Ghana require training to handle the array of very difficult questions posed by African Americans, such as those regarding the role of African tribes themselves in rounding up and selling Africans, which constituted an important function in the trans-Atlantic slave trade. Management at slave heritage sites must also anticipate the motivations and the emotions of African American visitors (Timothy and Teye 2004). This subgroup sometimes arrives in a mixed-race group tour and may undergo intense emotional experiences that transform their attitudes toward white tour group members. Indeed, they may potentially become downright hostile at the slave heritage site and/or on the journey from the site.

Given the sheer size of the continent and the complex integrated nature of heritage resources across individual countries, a regional approach should supplement those of individual countries. Organizations such as the Regional Tourism Organization of Southern Africa (RETOSA) and the Economic Community of West African States (ECOWAS) are likely to play greater roles in heritage tourism development in subSaharan Africa.

References

Ahmed, Z.U. (1992) Islamic pilgrimage (Hajj) to Ka'aba in Makkah (Saudi Arabia): an important international tourist activity. *Journal of Tourism Studies*, 3(1): 35–43.

Appiah, K.A. and Gates, Jr., H.L. (1999) *Encarta Africana: Comprehensive Encyclopedia of Black History and Culture*. Redmond, WA: Microsoft Corporation.

Austin, N.K. (1997) The management of historical sites of emotional significance to the visitor: the case of the Cape Coast Castle, Ghana. Unpublished PhD Thesis. Glasgow: Strathclyde University.

—— (2000) Tourism and the transatlantic slave trade: some issues and reflections. In P. U.C. Dieke (ed.), pp. 208–16. *The Political Economy of Tourism Development in Africa*. New York: Cognizant Communications.

Best, A.C.G. and de Blij, H.J. (1977) *African Survey*. New York: John Wiley.

Bruner, E.M. (1996) Tourism in Ghana – the representation of slavery and the return of the Black Diaspora. *American Anthropologist*, 98: 290–304.

Chevannes, B. (1994) *Rastafari: Roots and Ideology*. Syracuse, NY: University Press.

Curtin, P., Feierman, S., Thompson, L. and Vansina, J. (1995) *African History: From Earliest Times to Independence*. New York: Longman.

Ehert, C. (2002) *The Civilizations of Africa: A History to 1800*. Charlottesville, VA: The University Press of Virginia.

Essah, P. (2001) Slavery, heritage and tourism in Ghana. *International Journal of Hospitality and Tourism Administration*, 2(3/4): 31–49.

Garfield, R. (1994) *The Concise History of Africa*. Acton, MA: Copley Publishing Group.

Goodrich, J.N. (1985) Black American tourists: some research findings. *Journal of Travel Research*, 24(2): 27–28.

Haley, A. (1976) *Roots: The Saga of an American Family*. Garden City, NY: Doubleday.

Harris, J.E. (ed.) (1982). *Global Dimensions of the African Diaspora*. Washington, DC: Howard University Press.

Huggins, W.N. and Jackson, J.G. (1969) *An Introduction to African Civilizations, with Main Currents in Ethiopian History*. New York: Negro Universities Press.

Knight, G.C. and Newman, J.L. (1976) *Contemporary Africa: Geography and Change*. Englewood Cliffs, NJ: Prentice-Hall.

Malveaux, J. (1998) "Black power" means economic clout today. *USA Today*, February 6: 11A.

Mazrui, A. (1986) *The Africans: A Triple Heritage*. London: BBC Publications.

Ouma, J.P. (1970) *Evolution of Tourism in East Africa*. Nairobi: East Africa Literature Bureau.

Prentice, R. (1993) *Tourism and Heritage Attraction*. London: Routledge.

Shackley, M. (1999) Tourism and the management of cultural resources in the Pays Dogon, Mali. *International Journal of Heritage Studies*, 3(1): 17–27.

Teye, V. (1988) Coups d'etat and African tourism: a study of Ghana. *Annals of Tourism Research*, 15(3): 329–56.

—— (2008) International involvement in a regional development scheme: laying the foundation for national tourism development in Ghana. *Tourism Review International*, 12(3): forthcoming.

Teye, V., and Timothy, D. (2004) The varied colors of slave heritage in West Africa: white American stakeholders. *Space and Culture*, 7(2): 145–55

Teye, V., Sonmez, S.F. and Sirakaya, E. (2002) Residents' attitudes toward tourism development in Ghana: comparison of two cities. *Annals of Tourism Research*, 29 (3): 668–88.

Timothy, D.J. and Iverson, T. (2006) Tourism and Islam: considerations of culture and duty. In D.J. Timothy and D.H. Olsen (eds), *Tourism, Religion and Spiritual Journeys*, pp. 186–205. London: Routledge.

Timothy, D.J. and Olsen, D.H. (eds) (2006) *Tourism, Religion and Spiritual Journeys*. London: Routledge.

Timothy, D.J. and Teye, V.B. (2004) American children of the African diaspora: journeys to the motherland. In T. Coles and D.J. Timothy (eds), *Tourism, Diasporas and Space*, pp. 111–23. London and New York: Routledge.

Tunbridge, M. (1997) Whose heritage? Global problem, European nightmare. In G.J. Ashworth and P.J. Larkham (eds), *Building a New Heritage: Tourism, Culture and Identity in the New Europe*. London: Routledge.

UNDP (2008) *Annual Report*. New York: United Nations Development Program.

UNESCO (2008a) World Heritage List. Available from http://whc.unesco.org/en/list (accessed August 15, 2008).

—— (2008b) World Heritage List. Available from http://whc.unesco.org/en/list/35 (accessed August 16, 2008).

—— (2008c) World Heritage List. Available from http://whc.unesco.org/en/list/516 (accessed August 16, 2008).

—— (2008d) World Heritage List. Available from http://whc.unesco.org/en/list/119 (accessed August 16, 2008).

—— (2008e) World Heritage List. Available from http://whc.unesco.org/en/list/1139 (accessed August 16, 2008).

—— (2008f) *The Return of the Aksum Obelisk*. Paris: UNESCO.

UNWTO (2007) *Tourism Highlights, 2007 Edition*. Madrid: UN World Tourism Organization.

World Bank (2007a) *World Development Report: Development and the Next Generation*. Washington, DC: World Bank.

—— (2007b) *Africa Development Indicators*. Washington, DC: World Bank.

—— (2008) *World Development Report: Development and the Next Generation*. Washington, DC: World Bank.

WTO (1995) *National Tourism Development Plan for Ghana 1996–2010*. Accra: Ministry of Tourism, UNDP and WTO; Integrated Tourism Development Program (GHA/ 92/013).

—— (2000) *Tourism Market Trends*. Madrid: World Tourism Organization.

—— (2004) *WTO in Africa 1996–2003*. Madrid: World Tourism Organization.

—— (2005) *Overview of Tourism Performance in Africa*. Madrid: World Tourism Organization.

WTO/UNESCO (1995) *Accra Declaration on the WTO–UNESCO Cultural Tourism Programme*. Madrid: WTO/UNESCO.

11 Heritage management and tourism in the Caribbean

Leslie-Ann Jordan and David T. Duval

Introduction

The purpose of this chapter is to provide a snapshot of current regulatory and policy environments relevant to the development and maintenance of built, natural, and cultural heritage in the Caribbean (Figure 11.1). For clarity, the Caribbean is defined geographically as the string of small island states from Jamaica to Trinidad and Tobago, including Guyana and Venezuela in South America. These South American countries are often included in discussions of Caribbean issues owing to their location, physical geography, and physiographic connections to the Caribbean Basin, and their cultural and colonial similarities. Thus, while Chapter 12 addresses issues in Latin America, including some that span all of South America, Guyana is discussed in greater detail in this chapter. In this sense, some of the more common designations of "the Caribbean" as including Central American countries such as Belize or Honduras have been omitted. The scene of conquest, colonialism, and variable development throughout its history, the Caribbean offers an excellent laboratory within which issues such as government cooperation, supranational oversight, and international policy linkages can be viewed in the context of tourism development initiatives (Hall 2000; Timothy 2004). Geographically, it is immense, yet it features numerous independent states, which in some cases operate coherently and in concert with one another on some issues, yet independent of one another in others. The intent is not to provide a road map of government or even regional policy issues, but rather to highlight salient efforts to bring heritage and tourism together (Figure 11.1).

Tourism has emerged as one of the world's leading industries and the most important industry to the majority of the Caribbean countries (WTTC 2004). According to the World Travel and Tourism Council (WTTC 2004), tourism accounts for 15 percent of total Caribbean gross domestic product (GDP) and employment and about 20 percent of exports and investment. In 2005, the Caribbean received about 42.3 million visitors, of which 19.8 million were cruise ship visitors (CTO 2006). The region remains the premier cruising destination, accounting for roughly 48 percent of global cruise bed days over the last five years (WTTC 2004). In 2002, the Caribbean Tourism Organization

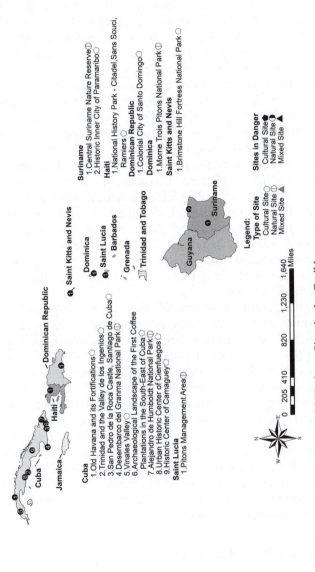

Cuba
1. Old Havana and its Fortifications ○
2. Trinidad and the Valley de los Ingenios ○
3. San Pedro de la Roca Castle, Santiago de Cuba ○
4. Desembarco del Granma National Park ⊕
5. Viñales Valley ○
6. Archaeological Landscape of the First Coffee
   Plantations in the South-East of Cuba ○
7. Alejandro de Humboldt National Park ⊕
8. Urban Historic Center of Cienfuegos ○
9. Historic Center of Camagüey ○

Saint Lucia
1. Pitons Management Area ⊕

Suriname
1. Central Suriname Nature Reserve ⊕
2. Historic Inner City of Paramaribo ○

Haiti
1. National History Park - Citadel, Sans Souci,
   Ramiers ○

Dominican Republic
1. Colonial City of Santo Domingo ○

Dominica
1. Morne Trois Pitons National Park ⊕

Saint Kitts and Nevis
1. Brimstone Hill Fortress National Park ○

Legend:
Type of Site
Cultural Site ○
Natural Site ⊕
Mixed Site ▲

Sites in Danger
Cultural Site ●
Natural Site ◉
Mixed Site ▲

0 205 410   820   1,230   1,640
                                  Miles

Figure 11.1 Developing countries and World Heritage Sites in the Caribbean

(CTO) estimated that more than 900,000 people were directly or indirectly employed in tourism in the Caribbean (CTO 2002). As such, the Caribbean is considered the most tourism-dependent region in the world, and tourism is the economic lifeblood of many Caribbean countries (WTTC 2004).

Tourism in the Caribbean has long been an enigma, and seen as either a blight or a necessary evil. Indeed, some (e.g., Best and Levitt 1975; Bryden 1973; Wilkinson 1997) have associated tourism with the dependency felt through centuries of colonial activity, while others have associated tourism development with engendering cultural decline (e.g., Britton 1989; Erisman 1983; Pattullo 1996) or the problematic nature of tourism as a means of development but at the risk of damaging sensitive resources (e.g., Grandoit 2005; Holder 1988). There is no question that the region has benefited from its proximal location to key markets such as the United States and Canada, and has welcomed the advent of modern jet engine aircraft, which reduced travel times from not only these markets but European ones as well (Duval 2004a). The region is now facing some of the more daunting tasks that other mature destinations have faced in recent decades: the question of improving yields, hanging on to tentative and often fickle markets, and improving the overall performance of the sector. Coupled with these challenges are the environmental concerns facing many island states, whether directly associated with tourism or not. It can even be argued that the public awareness and attention devoted to environmental issues relating to tourism (e.g., the carbon footprint of air travel) is enough to compound the issue further and thus bring the issue to the forefront of government planning.

To be fair, the Caribbean has arguably been a leader (regionally, at the very least) in environmental conservation and preservation, largely in association with "alternative" (now arguably the "mainstream") manner of tourism development (e.g., Duval 1998; Weaver 1995). This is not surprising given the fact that some of the largest groups of indigenous people in this hemisphere reside in two Caribbean states, namely Dominica and Guyana. More recently, there has been a growing awareness within many Caribbean islands of the need to redefine the tourism product, with a greater emphasis on sustainable development and community empowerment (Wilkinson 1989). Many Caribbean destinations have realized that their future tourism potential is almost solely based on the utilization of their rich natural and cultural heritage. In this regard, many island states have been embracing many different forms of sustainable tourism such as eco-tourism, nature tourism, soft-adventure tourism, and heritage/cultural tourism.

Heritage tourism in the Caribbean

Despite the region's geographic and demographic distinctions, the Caribbean has a rich common heritage, molded by slavery, colonialism, and the plantation (Pattullo, 1996). As Mather and Todd (1993: 9) document, most islands in the region have been under the influence of colonial powers at some time in their history, and the vernacular languages that are still spoken in the region reflect this history: French in Guadeloupe, Martinique, St. Martin, and Haiti; Dutch in Aruba, Bonaire, Curacao, St. Maarten, Saba, and St. Eustatius; Spanish in Cuba and the Dominican Republic; and English nearly everywhere else. The Caribbean's variety is regarded as one of its major selling points: its people, who are a mix of races; its different languages; its architecture and fortifications arising from its colonial past; its variety of religions ranging from Christianity to Hinduism to Voodoo; its varied topography, from the flat sand islands of the Bahamas to the lush tropical mountains of St. Lucia; and its cuisine, ceremonies, and traditions (Mather and Todd 1993: 139).

Unfortunately, specific data on heritage tourism have not been collected in the Caribbean, perhaps because only recently has heritage become an important issue in the region (Cameron and Gatewood 2008; UNESCO World Heritage Centre 2005) and because the industry has focused almost exclusively on sun, sea, and sand beach-based tourism. According to Honychurch (2005: 28), "The indigenous heritage of the Caribbean was almost literally wiped from the face of the earth during the violent conquest of the islands in the sixteenth century." He contends that, because of the loss of so much of the region's indigenous heritage, "it is the colonial and Creole elements of regional history that come to the fore where heritage and sustainable development are concerned, concentrated around our architectural heritage" (Honychurch 2005: 28). Given the region's history of colonization, slavery, indentureship, and struggles for independence, Pattullo (1996: 179) argues that:

> The most recent struggle for the Caribbean has been both to nurture its indigenous art forms, to create and perform for its own peoples, amid the demands of tourism, while at the same time finding imaginative ways of "using" tourists as patrons rather than being used by them.

There are also issues of concern relating to authenticity and cultural dependency, broad issues that are not recent (e.g., Harrigan 1974) but nonetheless pertinent to understanding the link between history, heritage, and tourism.

Another reason for the general neglect of heritage tourism is the fact that " ... the Caribbean tourism industry does not depend on castles, ancient buildings, art galleries and museums. The Caribbean's cultural forms are not on display as they are in Venice or Prague, Delhi or Cairo" (Pattullo 1996: 181). However, over the past five to ten years, in light of growing competition and the demand for an authentic experience (Timothy and Boyd 2006), there is a visible fundamental shift toward "alternative" forms of tourism, which is a departure from the region's traditional reliance on sun, sea, and sand. There is growing recognition and acceptance that the Caribbean has immense cultural and natural heritage tourism assets on account of its particular historical development and specific geographical and climatic conditions, which reflect the mixture of Amerindian, European, African, Asian, and other peoples (UNESCO World Heritage Centre 2005: 101). As a result, a magnificent ensemble of natural and archeological sites, cultural landscapes, historic towns and buildings, maritime heritage, as well as art works and traditions can be developed to form part of the tourism product.

At the conference on the Development of a Caribbean Action Plan in World Heritage in 2004, many Caribbean islands agreed that:

> our ability to survive as Caribbean and Small Island Developing States (SIDS) will depend on developing a new paradigm which is driven by strategies that take into consideration our diverse natural and cultural resources, our inspiring landscapes, our climate, our unique identity and the resilience and creativity of our people who have overcome centuries of hardship and exploitation ...
>
> van Hooff *et al.* (2005: 108)

As evidence of this new direction, the Fifth Annual Caribbean Conference on Sustainable Tourism Development, hosted by the Caribbean Tourism Organization (CTO) and the Association of Caribbean States (ACS) in 2003, was entitled "Keeping the Right Balance—Embracing our Heritage in the Greater Caribbean." In justifying the importance of the conference theme, Jean Holder, then the Secretary-General of the CTO, argued that:

> ... the number of those persons and those programmes focused on preserving, conserving and developing the very fundamentals on which the entire industry rests, our natural and built environment, our heritage and cultural identity, remain miniscule, in comparison to those engaged in generating the traffic which has the potential to destroy the very resources that make tourism possible. We must therefore continue to pay special attention to these issues and to hope that we can enlarge the number of people who see the value of what we are doing.
>
> Holder (2003: np)

Under the Global Strategy for a Balanced, Representative and Credible World Heritage List, adopted by the World Heritage Committee in 1994, the Caribbean region was earmarked as being under-represented in terms of the wealth and diversity of its natural and cultural heritage. In response, the World Heritage Center designed an Action Plan to assist Caribbean States Parties in the identification, protection, and conservation of their heritage and to provide financial and technical support to build capacity. Out of these initiatives, there are now six natural sites and twelve cultural sites, located in eleven Caribbean States Parties and Overseas Territories (see Table 11.1 and Figure 11.1). Also, fifteen Caribbean countries have ratified the World Heritage Convention (Table 11.2).

The following discussion will highlight specific facets of the tourism–heritage nexus in the Caribbean, some of which, as is shown, are closely linked with policy communities in the region relating to environmental performance,

Table 11.1 Selected UNESCO World Heritage Sites in the Caribbean

Country	World Heritage Site	Inscription date
Dominica	• Morne Trois Pitons National Park (N)	1997
St. Kitts and Nevis	• Brimstone Hill Fortress National Park (C)	1999
	• City of Charlestown (C)	Tentative list
	• Historic zone of Basseterre (C)	Tentative list
Barbados	• Bridgetown and its Garrison (C)	Tentative list
	• The Industrial heritage of Barbados: the story of sugar (C)	Tentative list
	• The Scotland district of Barbados (N)	
Grenada	• Grenadines Island Group (N)	Tentative list
	• St. George fortified system (C)	Tentative list
	• St. George historic district (C)	Tentative list
Jamaica	• Blue and John Crow Mountains National Park (CN)	Tentative list
St. Lucia	• Pitons management area (N)	2004
Guyana	• City Hall, Georgetown (C)	Tentative list
	• Fort Zeelandia (C)	Tentative list
	• Georgetown's plantation structure and historic buildings (C)	Tentative list
	• Shell Beach (Almond Beach) Essequibo Coast (N)	Tentative list
	• St. George's Anglican cathedral (C)	

Source: UNESCO World Heritage Centre (2007).

Table 11.2 Caribbean states party to the World Heritage Convention (as at October 2006)

	Year of adherence
Antigua and Barbuda	1983
Barbados	2002
Belize	1990
Cuba	1981
Dominica	1995
Dominican Republic	1985
Grenada	1998
Guyana	1977
Haiti	1980
Jamaica	1983
St. Kitts and Nevis	1986
St. Lucia	1991
St. Vincent and the Grenadines	2003
Suriname	1997
Trinidad and Tobago	2005

Source: UNESCO World Heritage Centre (2007).

visitor management, and overall tourism development. The intent is not to provide examples of best practice, largely because what constitutes best practices in one context may not be seen as such universally. The varied governmental and regulatory environment in the Caribbean bears this out; differing governmental approaches and levels of importance feature.

Natural heritage

Tourism activities in the Caribbean are largely associated with coastal zones. Often, these zones are the sites of significant environmental damage to the natural heritage of the location. Examples include irreparable damage to coral reefs from cruise ship anchors (e.g., Allen 1992), coastal erosion from over-development, and direct pollution and treatment. The World Travel and Tourism Council (2004) recently highlighted the problem of pollution and environmental degradation in its report on the impact of tourism on jobs and the economy in the region. It suggested that, overall, the success of tourism is largely dependent on maintaining the very natural environment on which it depends. Further, the WTTC (2004: 9) also highlighted the need for increased awareness of several issues including:

- Improved planning and management to increase the technical expertise required in the areas of pollution monitoring, coastal zone management, and the preparation and evaluation of Environmental Impact Assessments;
- Increased regional cooperation and collaboration;
- Higher standards of environmental quality;
- Conservation and the sustainable use of natural resources through participation in environmental certification and rating programs;
- Improvements in infrastructure across the region, notably in utilities such as water and electricity supply, and solid waste disposal;
- Greater clarity in land-use policy, containment of the spiraling price of land, and better zoning on the basis of maximizing economic returns;
- Incentives to mobilize the private sector to invest in environmental improvements;
- Education and in-service training for a more sustainable approach to tourism;
- Crisis and disaster management to mitigate the severe risk of natural and environmental disasters.

Naturally, these considerations require significant policy and planning measures. Several non-governmental organizations (NGOs) in the region are dedicated specifically to this purpose, including Environmental Protection in the Caribbean (EPIC) and the Island Resources Foundation (IRF). Government oversight of protection of natural environments is generally centered at the local level, although there have been examples of multilateral efforts. The Organization of Eastern Caribbean States (OECS) received some US$3.7

million in funding from the Global Environmental Facility[1] designed to "curb environmental degradation by strengthening the application of environmental safeguards and management capacity in six OECS member countries" (World Bank 2006: np). Indeed, the OECS established the Environment and Sustainable Development Unit (EDSU) (formerly the Natural Resources Management Unit) in 1986 with the specific purpose of the coordination of natural resource management throughout the Eastern Caribbean. In April 2001, the St. Georges Declaration of Principles for Environmental Sustainability in the OECS (SGD) was announced, consisting of twenty-one guiding principles relating to interactions with the natural environment. The basis of the agreement is that signatories to the Declaration (which includes all members of the OECS) are meant to adhere to these principles and seek advice and guidance from the OECS Secretariat where necessary (Table 11.3).

Built heritage

In 1994, the UNESCO World Heritage Committee's Global Strategy for a representative and balanced World Heritage List stated that the Caribbean region was under-represented in terms of the wealth and diversity of its natural and cultural heritage. Consequently, an action plan for the region was developed and, within that plan, four categories of cultural heritage were identified as being particularly significant for the Caribbean: fortifications, plantation systems, wooden heritage, and archeological sites (UNESCO World Heritage

Table 11.3 The principles of the St. Georges Declaration (SGD), announced 2001

1.	Better quality of life for all
2.	Integrated development planning
3.	More effective laws and institutions
4.	Civil society participation in decision-making
5.	Meaningful participation by the private sector
6.	Economic benefits from environmental management
7.	Broad-based environment education and awareness
8.	Preparation for climate change
9.	Integrated disaster management
10.	Preventing air, water, and land pollution
11.	Using available resources wisely
12.	Protecting natural and cultural heritage
13.	Protecting plant and animal species
14.	Sensible and sustainable trade
15.	Cooperation in science and technology
16.	Using energy efficiently
17.	Joint decision-making on international environmental agreements
18.	Coordinated work with the international community
19.	Putting the principles to work
20.	Obligations of member states
21.	Review and updating of the principles

Source: OECS (2003).

Centre 2005: 8). These four categories represent significant aspects of the region's built heritage that have been earmarked for identification, protection, and conservation. As such, between 1996 and 2004, several meetings, workshops, and conferences were held, organized by the World Heritage Centre in the framework of the Global Strategy Action Plan for the Caribbean. For example, at the Ninth Forum of Ministers of Cultural Heritage of Latin America and the Caribbean in 1997, which took place in Colombia, UNESCO formally endorsed the conclusions and recommendations of an expert meeting, which called for the nomination of a multinational "Fortifications in the Caribbean." The ministers decided to support the initiative of Colombia to work toward the inscription of a coherent ensemble of fortifications in the Caribbean on the World Heritage List. The report included a preliminary inventory of the most representative fortifications in the Caribbean, subdivided into four categories: fortified cities (e.g., Brimstone Hill Fortress National Park in St. Kitts and Nevis); garrisons (e.g., Bridgetown and its garrison in Barbados); military forts; and fortified systems (e.g., St. George Fortified System in Grenada) (UNESCO 1997; UNESCO World Heritage Centre 2005).

Another meeting on Wooden Urban Heritage in the Caribbean Region,[2] which was held in Guyana in 2003, originated " ... from the clear recognition of the vulnerability of the historic wooden architecture remaining in cities and towns of individual Caribbean islands today and the urgency of documenting and preserving this heritage, which is disappearing rapidly due to socio-economic changes, natural disasters, lack of maintenance and neglect" (UNESCO World Heritage Centre 2005: np). In light of these meetings, it is now left up to the individual Caribbean countries to translate debates and recommendations into national heritage inventory lists, tentative lists and, eventually, nominations of sites of potential outstanding universal value. Unfortunately, low priority is still given to cultural development and heritage preservation in terms of the attitudes of government officials, legislators, community and business leaders, whether on financial support for programs or the conception of cultural heritage as an integral part of sustainable development policies and programs (UNESCO 2005). Consequently, legislation to protect built heritage is uneven throughout the region, resulting in some countries having strict laws designed to protect their built heritage, while others have few, if any, regulatory frameworks.

St. Lucia has been one of the forerunners in terms of developing a regulatory system to preserve and protect the island's heritage assets. The St. Lucia National Trust, which was created by an Act of Parliament in 1975, has the legal mandate to conserve the natural and cultural heritage of the nation. Its efforts have received full government support and, as such, Section 33 of the Physical Planning Act No. 29 of 2001 specifically addresses the preservation of sites and buildings of interest. It states that the head of the Physical Planning and Development Division shall compile a list of buildings of special historic and/or architectural interest or may adopt, with or without modifications, any such list compiled by the National Trust as outlined in the 1975 Act (Marquis 2005).

In contrast, in Trinidad and Tobago, it took a long time and a major effort for the protection of heritage to achieve recognition. A National Trust Committee was set up in 1991 to prepare a draft of the National Trust Act of Trinidad and Tobago. The Committee suggested a legal framework for the preservation and conservation of the natural and built heritage of Trinidad and Tobago, as well as the establishment and *modus operandi* of the National Trust Council. However, despite great expectations, several years elapsed before the Citizens for Conservation could persuade the government to pursue the project seriously, and the National Trust of Trinidad and Tobago Act was finally passed in 1999. However, the Trust was still not set up for another eight years, during which time significant examples of the architectural heritage would be lost despite the efforts of Citizens for Conservation, among others (Lewis 2005: 77).

To help identify and preserve the region's built heritage, the Organization of the Wider Caribbean on Monuments and Sites (CARIMOS), which was created in 1982, has as its mandate to identify and study the historic monuments of the Caribbean region, while also providing technical assistance in the restoration and preservation of the built cultural heritage of the region. Over the years, CARIMOS has been actively working on establishing a Caribbean Heritage Database and, in 2002, the first phase of the Caribbean Cultural Heritage Inventory project was completed. The database includes approximately 1,000 monuments and sites in the fifteen African, Caribbean, and Pacific member countries.

Cultural heritage

Despite the fact that many island states in the region do not have direct, non-stop flights to major urban centers in North America or Europe, the scope of tourism development has had some impact on the kinds of visitor experiences constructed. In many respects, the path of tourism development, in both a spatial and a temporal sense, has followed a similar trajectory to colonialism. Overall, however, the question remains as to what effect tourism has had on cultural heritage in the region. In one sense, the June 2007 Conference on the Caribbean held in Washington, DC, to celebrate American–Caribbean Heritage Month demonstrates the attention given to matters of cultural heritage. For this particular conference, these matters were said to be of particular importance given the position of the Caribbean in wider global trade and investment flows, and the intent of the conference was to solidify the relationship between CARICOM (Caribbean Community) and the United States, particularly with reference to the Millennium Development Goals. Heritage was not overtly a target, but provision was made for the integration of cultural tourism within wider partnerships and cooperation agreements.

In the region, cultural heritage has wide meaning and even wider application. It can have reference to historic places and buildings or social and cultural practices (e.g., the history of soca music), both endemic to the region or as a result of forced migration. As indicated above, direct policy on the management of social and cultural heritage is generally relegated to individual island states, unlike cross-country agreements on built and/or natural heritage (discussed above/below). Supraregional efforts, however, are evident and numerous, although not always binding in a regulatory manner. For example, an agreement forged between CARICOM and UNESCO in Guyana in 2003 solidified the dedication of both toward preserving the cultural heritage of the region:

> CARICOM and UNESCO will cooperate in safeguarding the tangible and intangible cultural heritage of the Caribbean through ratification and implementation of Conventions on World Heritage and Underwater Cultural Heritage, in support of intercultural heritage and cultural diversity, in jointly developing cultural enterprises in the Region, through support for the development and implementation of cultural policy at regional and national levels, and through integration of cultural approaches in addressing pressing regional concerns such as youth development, HIV/AIDS prevention, drug abuse reduction and the promotion of peace.
>
> CARICOM (2003)

This partnership between UNESCO and CARICOM has led to a series of capacity development activities held in the Caribbean to promote the digital preservation of cultural heritage in the region. For example, the first Regional Workshop on Digitization of Cultural Heritage in the Caribbean and Training in UNESCO's Ibero-American Digital Library Software, organized by UNESCO in collaboration with the ICT4D Jamaica, Human Education Art and Resource Training/National Training Agency (HEART/NTA), International Institute for Communication Development (IICD), and Institute for Connectivity of the Americas (ICA), was held in 2005 in Jamaica. Over forty representatives from the region participated. This workshop, along with other sessions held in the region, has contributed to the enhanced capacity of specialists who operate within public, private, and NGO entities, such as libraries, museums, archives, national cultural commissions, and producers of cultural content, to deal with aspects of the digital preservation and documentation of cultural heritage (UNESCO 2005).

Further, the OECS, in an agreement with France (given that France administers Overseas Departments in the Caribbean such as Martinique and Guadeloupe), established an online network (OECSCulture.net) to facilitate the networking of cultural groups throughout the region. The variability of wider regional cooperation in heritage tourism matters has met problematic and uneven funding patterns. The Organization of American States (OAS) reviewed the amount of funding available for heritage and cultural tourism and discovered that, while readily available, sources for funding were not always available on an equal basis (Organization of American States 1995). Indeed, the Organization of American States (1995) proposed several reforms for governments in the region to enhance heritage tourism development.

First, national plans should make clear the extent of desired natural and heritage tourism growth and should be viable and feasible from a natural resource as well as an administrative perspective. Second, private investment in heritage and nature tourism is critical and should be incentivized and encourage through easier access to funding and credit. The third reform is that the question of control of natural and heritage resources needs to be resolved, and governments may need to cede some control in instances where private capital investment will bring positive returns on investment. Fourth, related to point three, preservation and conservation need not only be under government purview, particularly when various NGOs exhibit the necessary skills and desire to effectively manage. Government oversight can still occur, but "on-the-ground" management may not always be most effective coming from the government itself. Fifth, an environmental tax, however levied, should be seen as dedicated revenue for the enhancement of production of dedicated heritage tourism projects rather than used as direct general revenue. Finally, residents can be seen to be ambassadors and sources of information about natural, built, and cultural heritage, but only if they are involved in the process and educated about their significance.

Case studies: Caribbean heritage management issues

Having identified some of the wider circumstances through which tourism and heritage feature within supraregional and/or national policy directives, the following case studies highlight some of the issues and challenges that three Caribbean countries have been encountering as they seek to develop and manage heritage tourism. Countries such as Guyana, Dominica, and Jamaica have incorporated the development and management of heritage and cultural tourism as major components in their national development strategies. They have been highlighting varying aspects of their built, natural, and cultural heritage.

Guyana

Guyana is the poorest CARICOM economy in terms of per capita income and is considered the least developed in terms of physical and social infrastructure (Khan 2006). Given the fact that Guyana is rich in natural resources, natural sites, and flora and fauna, Guyana's perceived future is in nature-based, adventure, and heritage and cultural tourism. The government of Guyana, in its *National Development Strategy 2001–2010*, has stated that heritage and cultural tourism can promote a number of Guyana's sites for both their historical and their architectural value. Sites suitable for this type of tourism include Georgetown, Fort Island, Magdelenburg and Kyk-Over-Al, and Kaieteur Falls. The *Strategy* has outlined several initiatives that would need to be developed to highlight the country's heritage product, while minimizing the negative impact of tourism development. One of these initiatives involves the creation of a Protected Area System or, at the very least, the according of special status to areas known to possess unique natural characteristics. Additionally, there are plans to enforce strict zoning and building codes in the capital, Georgetown. New buildings in the city will hopefully be made to conform to Georgetown's rich architectural heritage, as sections of the city represent significant opportunities for architectural preservation and the development of tourism sites. It is proposed that building in the city be very carefully regulated and monitored to preserve the product before it is completely devastated by new developments. The intention is also to encourage investors by offering them tax incentives to develop small-scale inns in the style of the existing historic architecture of Georgetown (Government of Guyana 2005a).

The government recognizes that, although Amerindian, African, and Indian cultures are of potential interest to tourists, it is important to protect these cultures and communities, particularly the indigenous Amerindian communities, from the negative impacts that tourism can have on their traditional ways of life. For example, the *Strategy* states:

> The influence of foreign cultures may also impact upon communities in such a way that traditional values may be lost. Moreover, the commercialization of culture can lead to the development of a pseudo-culture and folklore that have been specially devised for tourists, the alteration of traditional crafts because of commercial pressures, and the replacement of traditional handicrafts by less authentic but more saleable souvenirs.
>
> Government of Guyana (2005a: np)

However, the government realizes that heritage tourism, "presents Amerindians with an opportunity to build an indigenous industry which is labour intensive and would benefit local communities" (Government of Guyana 2005a: np).

Subsequently, the *Strategy* outlines at length the measures and policies that would be implemented. The first of these measures is the involvement and support of Amerindian communities and other people living in the hinterland and is recognized as essential for the development of an effective park system that can significantly attract and support tourism. The level of this involvement should include planning and policy-making at the national and local levels. It is equally necessary that social partnerships are encouraged between Amerindians and private investors, and that Amerindian communities have access to capital to foster their own direct involvement in the industry.

The following is a summary of some additional policies that the government plans to implement:

- Tourist agencies and the National Protected Areas System, when it is established, will involve Amerindians in their eco/heritage tourism activities. Amerindians could be trained as park rangers and guides, as they have an unrivaled knowledge of the local terrain and its natural resources.
- There are plans for fundamental institutional strengthening, whereby a participatory approach is applied through direct discussion, education, and practical training programs. Amerindian groups should also be empowered

financially and otherwise to start their own tourist ventures in a small and manageable way.

• Amerindian communities will decide for themselves if heritage-based, nature-based, and eco-tourism ventures are worthy of their involvement and participation, on a project-by-project basis.

• Tourism activities will hopefully be started at a slow and measured pace in Amerindian communities so as not to overwhelm local capacity and result in an increase in social stresses (Government of Guyana 2005b).

Dominica

Dominica—The Nature Island of the Caribbean advertises its Carib community as one of its main cultural heritage products. Even the government indicates that "Caribs play a significant role in tourism through the marketing of their traditional handicrafts, and their reserve is an attraction for tourist visitors" (Government of Dominica 2006: 52). Dominica has the largest population of indigenous people in the Eastern Caribbean, approximately 3,000 or 4 percent of the population. About 1,700 resident Carib people live on a 3,700-acre reserve in the northeast of the island. Although the Carib community has an autonomous political structure and communal lands, Caribs are the most disadvantaged group in the country. Recent information on poverty in Dominica found that, " ... poverty in Dominica is high – around 29% of households and 39% of the population, which is high by Caribbean standards ... Poverty amongst the Caribs is much higher: 70% of the Carib population is poor and almost half are indigent" (Government of Dominica 2006: 7).

Living conditions in the Carib community are very poor by Caribbean and world standards. There is very little formal employment and few possibilities for other means of generating incomes, and social services workers indicate informally that there are significant problems with alcoholism, abuse, and overall low self-esteem. Few young people finish high school and train for careers in government or the private sector. Handicrafts (bags, baskets, etc.) produced by Carib artisans are sold to tourists who pass through the reserve. However, as Khan (2006: 24) stated, "it is unlikely that the tour buses carrying tourists would linger long in the area since the poor conditions may be upsetting to people who are on holiday. Tour buses laden with tourists from cruise ships drive through the reserve to places of interest further away." Several of the Carib homes are made of galvanized roofing material on all sides, which must be extremely uncomfortable in the heat and humidity. Others are very old wooden huts in need of repair (Government of Dominica 2006). The government has also reported that 39 percent do not have access to safe water, virtually none have a flush toilet, and fewer than 30 percent have proper kitchen facilities (Government of Dominica 2006: 7).

One of the major challenges facing heritage tourism, which is the source of development in the Carib community, is the fact that the natives have neither the capital nor the management skills or knowledge to operate tourism businesses (Khan 2006). There is one guest house on the reserve with six rooms, but the accommodation is basic at best and will not meet the standards of any foreign tour operators or average tourists (see http://www.avirtualdominica.com/ctgh.htm). There is no air conditioning, little privacy, and the overall appearance is unimpressive. Perhaps one or two younger adventure tourists or "backpackers" may stay at the guest house, but it is unlikely to attract mainstream tourists. It is a major challenge to find a strategy through which increased tourism will benefit the poorest group in Dominica. The Carib community needs major investment in infrastructure, education, and training to enable them to become players in the tourism industry. They will also need substantial capital investment to build proper accommodation and other facilities, but individuals apparently do not hold title to land which banks normally required for mortgages.

In its *Medium Term Growth and Social Protection Strategy*, the government of Dominica included a "Special Focus on the Carib Community." There is a Comprehensive Carib Territory Community Development Program to be financed by the Caribbean Development Bank (CDB) and the Dominican government (US$4.3 million) that covers the four-year period 2003–7. The program seeks to diversify the economy of the Carib Territory by providing opportunities for employment to improve the well-being of the indigenous people by undertaking investments in the tourism sector (Carib model village), health sector (construction of a health center), road infrastructure (feeder roads), agriculture and livestock, education and housing, land reform, community resource centers, and other social activities.

The government's medium-term plan indicates that it will implement a national tourism policy, but it does not clearly articulate a specific focus on tourism and poverty alleviation. The reference to tourism relates to land use planning (Government of Dominica 2006: 61), but it aims to promote cultural and heritage tourism. The stated contribution from tourism to heritage/cultural protection includes:

- Protection of heritage and cultural resources throughout Dominica with particular attention to community-based resources;
- Incorporation of heritage interpretive programs into the tourism product mix including Roseau water mills and other features of Dominica's heritage;
- Incorporation of community-based heritage and cultural products into scenic parkway programs;
- Generation of tourism revenues for Dominica's cultural enterprises and groups; and
- Fostering pride in, and support for, Dominica's culture (Government of Dominica 2006: 50).

In 2006, the Carib community opened Kalinago Barana Aute (or "Carib Village by the Sea") (http://www.kalinagobaranaaute.com), which is advertised as "a unique Caribbean facility intended to contribute in several ways to the socio-economic development of the Kalinago people of Dominica." Visitors are encouraged to sample the lifestyle of the Kalinago people and learn about their history through interpretive dance, language, and crafts. A typical visit to Kalinago Barana Aute includes the opportunity to learn traditional basket weaving, catch crayfish using traditional methods, and learn how the Kalinago Indians used the trunk of the gigantic Gommier tree to construct their most valuable possession, the canoe. Visitors will also be entertained by cultural performances, calabash art story telling, face painting, and a Kalinago lecture. Additionally, guests will have the opportunity to obtain a spiritual cleansing, visit the on-site herbal garden, and may even purchase herbal medicine at the conclusion of the tour (Dominica Tourist Board 2006).

The Caribs of Dominica are perhaps an excellent example of the reification of a cultural and social tradition through the lens of tourism. Indeed, the manager of the Carib Village said in a story in *USA Today* that foreign interest cannot help but strengthen traditions: "It's an opportunity to right some wrongs ... It sends the message to people here that there's something significant about them. If you can get people to appreciate what's unique about them, then their cultural heritage is in good hands." (Clark 2007: np). In nearby St. Vincent, heritage and tourism have been the focus of exploring the complex ethnohistory of the Black Carib, where some research (e.g., Duval 2004b) has suggested that complex relationships involving identity and tradition may explain efforts to enhance visitor experiences.

Jamaica

Jamaica has realized the need to make its tourism product more attractive to both domestic and international visitors by focusing on its heritage assets, as well as the traditional three Ss (sun, sea, and sand). Consequently, the Tourism Product Development Company Ltd of Jamaica (TPDCo) has been placing greater emphasis on the development of cultural heritage tourism to widen the market for eco-, nature-based, and adventure tourism. TPDCo believes that developing cultural heritage tourism can benefit Jamaican communities and the country at large by:

- Creating jobs and businesses;
- Diversifying the local economy;
- Creating opportunities for partnerships;
- Attracting visitors interested in history and preservation;
- Increasing historic attraction revenues;
- Preserving local traditions and culture;
- Generating local investment in historic resources;
- Building community pride in heritage (Tourism Product Development Company Ltd of Jamaica 2007: np).

As such, one of the specific objectives of its *Master Plan for Sustainable Tourism Development in Jamaica (2001–2010)* is "providing strategies for the development of tourism products and facilities in the context of environmental and cultural preservation and conservation." The plan has five main objectives, one of which is growth based on a sustainable market position. Under this objective, the plan notes that, for Jamaica to keep its tourism industry growing, it needs to hold its competitive space in the market by offering visitors what no other country has—its own unique Jamaican heritage. This heritage is defined as its natural surroundings, culture, history, and historic buildings and sites.

To improve visitors' heritage experience, the Ministry of Tourism plans to work with the heritage agencies on protection and conservation activities, making the development of the tourism heritage product a joint effort. There will also be concentration on four heritage sites with international appeal, Port Royal, Spanish Town, Falmouth, and Seville, supported by scenic routes and circuits and ten heritage trails across the country. Some of the themes for the heritage trails include natural wonders, slavery and emancipation (the Maroons), forts and fortifications, churches, great houses, industrial heritage, pre-Columbian heritage, and Jamaican culture and music.

The country also plans to use heritage tourism as the basis for urban renewal by encouraging community organizations to present their towns and villages as possible heritage resources for funding. A Heritage Challenge Fund is earmarked to be established to provide resources in the following ways:

- For towns and villages of culture, one town would be selected each year for a contribution of up to US$2 million for its development as a cultural/heritage center. The local community would provide at least 20 percent of the resources, including contributions from individuals and businesses in cash, goods, or services.
- For Tourism Development Action Plan Areas (TDAPAs), parts of towns or villages with strong tourism potential would be eligible for grants, technical assistance, or loans to develop facilities. The TDAPAs would be selected annually. They would get grants of up to US$500,000, with the local community providing 20 percent. Community-based ventures, heritage attractions, and private sector marketing and product development projects would qualify for different kinds and levels of assistance.
- It is proposed that the Heritage Challenge Fund be financed from several different sources including a special levy, a matching contribution from government, and international grants.
- To establish a Heritage Unit in the Tourism Product Development Co. Ltd (TPDCo) to administer the Heritage Fund, process applications, provide technical assistance, and monitor projects (*Master Plan for Sustainable Tourism Development in Jamaica 2001–10*).

Given that cooperation between agencies is an essential critical success factor for developing heritage tourism, TPDCo, along with the Ministry of Tourism, is working closely with the Jamaica National Heritage Trust, which was legalized in 1985. The primary functions of the Jamaica National Heritage Trust are:

- To promote the preservation of national monuments and anything designated as protected national heritage for the benefit of the island;
- To conduct such research as it thinks necessary or desirable for the purposes of the performance of its functions under the Jamaica National Heritage Act;
- To carry out such development as it considers necessary for the preservation of any national monuments or anything designated as protected national heritage;
- To record any precious objects or works of art to be preserved and to identify and record any species of botanical or animal life to be protected (Jamaica National Heritage Trust 2005)

Future challenges/critical success factors

The preceding case studies offer a snapshot of salient components of a strategic heritage tourism focus in some contexts. Beyond these specifics, however, several challenges can be identified for the development of heritage tourism in the Caribbean. Some of these include:

- Political will—Much of the planning and policy direction throughout the region is set by national governments, although at times under the auspices of international or regional agreements. Critically, the implementation and monitoring of projects relating to heritage preservation need to be seen as politically viable. While this is a difficult task in often unpredictable economic environments, it is nonetheless the backbone through which heritage tourism development (and, by extension, conservation) will proceed in the region.
- Marketing—Many countries worldwide have utilized heritage in marketing materials. Given its proximity to the large North American market, it will be critical for the message of heritage and opportunities for intellectually stimulating tourist activities to be communicated coherently. As Found (2004) has noted, the Caribbean has made significant steps toward successfully integrating tourism into its heritage product and heritage into its tourism product, and there is increased identification of the needs of further enhancements through funding. In many respects, the post-colonial market is strategically a viable and significant market for the region (Momsen 2004).
- Capital funding—As discussed, the lack of direct capital financing has hindered structured developments and preservation efforts. Part of this relates to government inability or unwillingness to inject the necessary

capital. From the private investment perspective, it needs to be shown whether projects involving heritage, conservation, and tourism will result in a positive return on investment (see, for example, Haywood and Jaya-wardena 2004). As a result, a positive investment climate must be fostered, including options such as interest-free loans, public–private partnership, and strategic joint venture agreements.

- Institutional arrangements—The importance of appropriate institutional arrangements for tourism development in small island developing states (SIDS) cannot be understated (Jordan 2004, 2007). Many of the national and regional organizations entrusted with the responsibility of developing heritage tourism are too underfunded and understaffed to carry out their responsibilities. Given this scenario, interorganizational partnerships become more important in sharing scarce financial and human resources.
- Community interest, involvement, and participation—Continuing to involve the local community in heritage management is critical (Sutty 1998). Such involvement will allow local residents to participate in the decision-making process that dictates what developments occur in their communities. It allows local people to participate in and take advantage of the economic opportunities that may occur through the provision of goods and services for heritage tourism, and it helps to empower residents and encourages them to take responsibility for managing their fragile resources (St. Lucia Heritage Tourism Programme 2002).

Conclusion

This chapter has attempted to encapsulate some of the more salient issues surrounding the heritage management of tourism resources in the Caribbean. In so doing, heritage was utilized in a broad manner, reflecting natural, built, and living cultural aspects. All of these permeate throughout Caribbean "society," and all have incredible significance for the future of Caribbean tourism. Two broad concluding statements can be made. The first is that heritage is often the very heart of the Caribbean (or at least "West Indian") visitor experience. This is, in large part, due to the marketing reality and shift in destination branding that sun, sand, and sea are no longer the only reason for visiting when various heritage sites and experiences enjoy success in the wider tourism sector. Thus, it is imperative that marketing efforts continue to focus on what are no longer "alternative" experiences: the cultural, natural, and built heritage of the region. Further to this is the development of meaningful, accessible, and directed strategies for smart development of appropriate product.

Second, regulation and preservation of heritage (broadly interpreted) needs a wider regional approach, but local efforts and knowledge may need to take precedence in some situations. Advocates of wider, top-level approaches to policy and planning in tourism may conveniently forget local knowledge production that, in some cases, may follow more environmentally, socially, and economically beneficial paths.

As already noted, there are two critical challenges facing heritage tourism management in the Caribbean. Rather than subsume the first challenge under the common rubric of "product development," in the Caribbean, it is the level of appropriate investment and development paths that illuminates the specific challenges at present. Investment and development paths will continue to require a balance between the economic contribution made by tourism activities, cultural sensitivities (and the risk of commoditizing cultural practice), and the strength of local/community input. The second critical challenge is wider economic stability in both the region and the generating markets. This is not related to heritage *per se*, but the general economic performance of SIDS in the region that can reflect on investment in and management of heritage resources, just as the economic performance of tourist source markets can dictate future visitation flows.

The Caribbean has changed considerably since the introduction of jet aircraft in the 1950s/1960s, which propelled the region in various markets (but primarily the United States) as a sun, sand, and sea regional getaway. Increasing competition and the resulting maturation of the product on offer have widened potential visitor experiences and extended (some might even argue solidified) the economic importance of tourism. Heritage tourism continues to play a critical role and, as this chapter has outlined, there are substantial efforts to harness the capacity for heritage experiences.

Notes

1 An independent organization whose projects are generally managed by one of the UNEP, UNDP, and the World Bank.
2 This meeting was organized by the UNESCO World Heritage Centre in cooperation with the Government of Guyana, the UNESCO Office for the Caribbean in Jamaica, and the Organization of the Wider Caribbean on Monuments and Sites.

References

Allen, W.H. (1992) Increased dangers to Caribbean marine ecosystems: cruise ship anchors and intensified tourism threaten reefs, *Bioscience* 42(5): 330–35.

Best, L. and Levitt, I. (1975) Character of the Caribbean economy. In G. Beckford (ed.),*Caribbean Economy: Dependence and Backwardness*, pp. 34–60. Mona: Institute for Social and Economic Research, University of the West Indies.

Britton, S. (1989) Tourism, dependency and development: a mode of analysis. In T. V. Singh, H.L. Theuns and F.M. Go (eds), *Towards Appropriate Tourism: The Case of Developing Countries*, pp. 155–72. Frankfurt am Main: Peter Lang.

Bryden, J.M. (1973) *Tourism and Development: A Case Study of the Commonwealth Caribbean*. London: Cambridge University Press.

Cameron, C.M. and Gatewood, J.B. (2008) Beyond sun, sand and sea: the emergent tourism programme in the Turks and Caicos Islands. *Journal of Heritage Tourism*, 3 (1): 55–73.

Caribbean Tourism Organization (CTO) (2002) *Caribbean Tourism Statistical Report*. Barbados, CTO.

—— (2006) Caribbean Tourism Performance 2005. Available from http://www.onecari bbean.org (accessed March 3, 2007).

CARICOM (Caribbean Community) (2003) Memorandum of understanding between the Caribbean Community (CARICOM) and the united Nations Educational, Scientific and Cultural Organisation (UNESCO), 5 May 2003, Georgetown, Guyana. Available from http://www.caricom.org/jsp/secretariat/legal_instruments/mou_carico m_unesco_03.jsp?menu = secretariat (accessed October 10, 2007).

Clark, J. (2007) The old ways still dominate in Dominica. *USA Today*, May 7. Available from http://www.usatoday.com/travel/destinations/2007-07-05-dominica_N.htm (accessed July 22, 2007).

Dominica Tourist Board (2006) Kalinago Barana Autê – the Carib cultural village by the sea. Available from http://www.dominica.dm/site/caribvillage.cfm (accessed July 28, 2007).

Duval, D.T. (1998) Alternative tourism on St. Vincent. *Caribbean Geography*, 9(1): 44–57.

—— (2004a) Trends and circumstances in Caribbean tourism. In D.T. Duval (ed.), *Tourism in the Caribbean: Trends, Development, Prospects*, pp. 3–22. London: Routledge.

—— (2004b) Cultural tourism in postcolonial environments: negotiating histories, ethnicities and authenticities in St. Vincent, Eastern Caribbean. In C.M. Hall and H. Tucker (eds), *Tourism and Postcolonialism: Contested Discourses, Identities and Representations*, pp. 57–75. London: Routledge.

Erisman, M (1983) Tourism and cultural dependency in the West Indies. *Annals of Tourism Research*, 10: 337–61.

Found, W.C. (2004) Historic sites, material culture and tourism in the Caribbean islands. In D.T. Duval (ed.), *Tourism in the Caribbean: Trends, Development, Prospects*, pp. 136–51. London: Routledge.

Government of Dominica (2006) *Medium-Term Growth and Social Protection Strategy*. Roseau: Government of Dominica. Available from http://siteresources.worldbank. org/INTPRS1/Resources/Dominica_PRSP(April2006).pdf (accessed July 22, 2007).

Government of Guyana (2005a) *Guyana's National Development Strategy 2001–2010*. Available from http://www.ndsguyana.org/default.html (accessed July 23, 2007).

—— (2005b) *Guyana Tourism Development Action Plan (2006–2010)*. Georgetown, Guyana: Ministry of Tourism, Industry and Commerce.

Government of Jamaica (2001) *Sustainable Tourism Master Plan Jamaica (2001–2010)*. Available from http://www.jsdnp.org.jm/susTourism-masterPlan.htm (accessed July 21, 2007).

Grandoit, J. (2005) Tourism as a development tool in the Caribbean and the environmental by-products: the stresses on small island resources and viable remedies. *Journal of Development and Social Transformation*, 2. Available from http://www. maxwell.syr.edu/moynihan/Programs/dev/journal.html (accessed December 14, 2008).

Hall, C.M. (2000) *Tourism Planning: Policies, Processes and Relationships*. Harlow: Prentice Hall.

Harrigan, N. (1974) The legacy of Caribbean history and tourism. *Annals of Tourism Research*, 2(1): 13–25.

Haywood, K.M. and Jayawardena, C. (2004) Tourism businesses in the Caribbean: operating realities. In D.T. Duval (ed.), *Tourism in the Caribbean: Trends, Development, Prospects*, pp. 218–34. London: Routledge.

Holder, J. (1988) Pattern and impact of tourism on the environment of the Caribbean. *Tourism Management*, 9:119–27.

—— (2003) CTO Secretary General's Address at Opening of STC-5. Available from http://www.1caribbean.org/information/documentview.php?rowid = 1695 (accessed August 7, 2007).

Honychurch, L. (2005) Caribbean heritage: its uses and economic potential. In Caribbean Wooden Treasures, Proceedings of the Thematic Expert Meeting on Wooden Urban Heritage in the Caribbean Region, UNESCO World Heritage Center. Available from http://whc.unesco.org/documents/publi_wh_papers_15_en.pdf (accessed August 10, 2007).

Jamaica National Heritage Trust (2005) Jamaica National Heritage Trust Act 1985. Available from http://www.jnht.com/act_1985.php (accessed August 10, 2007).

Jordan, L. (2004) Institutional arrangements for tourism in small-twin island states of the Caribbean. In D. Duval (ed.), *Tourism in the Caribbean: Trends, Development, Prospects*, pp. 99–118. London: Routledge.

—— (2007) Interorganisational relationships in small twin-island developing states in the Caribbean – the role of the internal core–periphery model: the case of Trinidad and Tobago. *Current Issues in Tourism*, 10(1):1–32.

Khan, P. (2006) *Challenges and Opportunities Presented by Growth of the Tourism Industry in the Caribbean: Implications for Poverty Alleviation*. Report prepared for the International Development Research Centre (IDRC), Montevideo, Uruguay.

Lewis, V. (2005) Protecting the wooden urban heritage in Trinidad and Tobago. In Caribbean Wooden Treasures, Proceedings of the Thematic Expert Meeting on Wooden Urban Heritage in the Caribbean Region, UNESCO World Heritage Center, pp. 75–78. Available from http://whc.unesco.org/documents/publi_wh_papers_15_en.pdf (accessed August 10, 2007).

Marquis, D. (2005) Saint Lucia's wooden urban heritage. In Caribbean Wooden Treasures, Proceedings of the Thematic Expert Meeting on Wooden Urban Heritage in the Caribbean Region, UNESCO World Heritage Center, pp. 71–73. Available from http://whc.unesco.org/documents/publi_wh_papers_15_en.pdf (accessed August 10, 2007).

Mather, S. and Todd, G. (1993) *Tourism in the Caribbean*. Special Report No. 455. London: Economist Intelligence Unit.

Millar, L. (2005) Tourist cruise ships and the trade in services: recent trends in countries of the Caribbean Basin. *Bulletin FAL*, 223. Available from http://www.eclac.cl/Transporte/noticias/bolfall/8/20968/FAL223.htm (accessed August 21, 2006).

Momsen, J.H. (2004) Post-colonial markets: new geographic spaces for tourism. In D. T. Duval (ed.), *Tourism in the Caribbean: Trends, Development, Prospects*, pp. 273–88. London: Routledge.

Organisation of American States (1995) The financing requirements of nature and heritage tourism in the Caribbean. Available from http://www.oas.org/dsd/publications/Unit/oea78e/ch08.htm (accessed July 29, 2007).

OECS (Organization of East Caribbean States) (2003) The St. Georges' Declaration of principles for environmental sustainability in the OECS. Available from http://www.oecs.org/esdu/sgd_Principles.html (accessed January 8, 2009).

Pattullo, P. (1996) *Last Resorts: The Cost of Tourism in the Caribbean*. London: Cassell.

St. Lucia Heritage Tourism Programme (2002) *Critical Factors for Market Success of Nature Heritage Tourism in Saint Lucia: A Guide for Planners and Developers*. St. Lucia: St. Lucia Heritage Tourism Programme.

Sutty, L. (1998) Local participation in tourism in the West Indian islands. In E. Laws, B. Faulkner and G. Moscardo (eds), *Embracing and Managing Change in Tourism: International Case Studies*, pp. 222–34. London: Routledge.

Timothy, D.J. (2004) Tourism and supranationalism in the Caribbean. In D.T. Duval (ed.), *Tourism in the Caribbean: Trends, Development, Prospects*, pp. 119–35. London: Routledge.

Timothy, D.J. and Boyd, S.J. (2006) Heritage tourism in the 21st century: valued traditions and new perspectives. *Journal of Heritage Tourism*, 1(1): 1–16.

Tourism Product Development Company Ltd of Jamaica (TPDCo) (2007) Cultural Heritage Tourism. Available from http://www.tpdco.org/dynaweb.dti?dynasection = tourismenhancement&dynapage = culturalt (accessed December 12, 2007).

UNESCO (1997) Ninth Forum of Ministers of Cultural Heritage of Latin America and the Caribbean endorses proposed WH nomination of "Fortifications in the Caribbean".

—— (2005) Promoting digital preservation of cultural heritage in the Caribbean. Available from http://portal.unesco.org/ci/en/ev.php-URL_ID = 20723&URL_DO = DO_TOPIC&URL_SECTION = 201.html (accessed July 1, 2007).

UNESCO World Heritage Centre (2005) Caribbean Wooden Treasures, Proceedings of the Thematic Expert Meeting on Wooden Urban Heritage in the Caribbean Region. Available from http://whc.unesco.org/documents/publi_wh_papers_15_en.pdf (accessed August 10, 2007).

—— (2007) World Heritage List. Available from http://whc.unesco.org/en/list (accessed August 5, 2007).

van Hooff, H., Armony, L. and van Oers, R. (2005) Conference on the Development of a Caribbean Action Plan, Castries, Saint Lucia, 2004. Summary Reports of Thematic Expert Meetings in the Caribbean, UNESCO World Heritage Center.

Weaver, D. (1995) Alternative tourism in Montserrat. *Tourism Management*, 16(8): 593–604.

Wilkinson, P.F. (1989) Strategies for tourism development in island microstates. *Annals of Tourism Research*, 16: 153–77.

—— (1997) Jamaican tourism: from dependency theory to a world-economy approach. In D.G. Lockhart and D. Drakakis-Smith (eds), *Island Tourism: Trends and Prospects*, pp. 181–204. London: Pinter.

World Bank (2006) Grant to eastern Caribbean states will boost environmental protection. Available from http://go.worldbank.org/3CNLYR0NF0 (accessed March 19, 2007).

World Travel and Tourism Council (WTTC) (2004) *The Caribbean: The Impact of Travel and Tourism on Jobs and the Economy*. London: World Travel and Tourism Council.

12 Heritage tourism in Latin America

Can turbulent times be overcome?

Regina Schlüter

Introduction

Latin America awoke the interest of travelers in the eighteenth and nineteenth centuries who, anxious to become familiar with the exotic flora and fauna of the region, ignored the discomforts and risks they had to face during their long voyages. The situation changed at the beginning of the nineteenth century, when local elites decided to build luxury hotels for their own use and for wealthy foreign visitors. Nevertheless, it was not until the 1960s that Latin America tried to achieve development through tourism either as an economic alternative to achieve growth or as a supplement to economic activities (Schlüter 1998). According to Getino (1993), tourism appeared as a kind of life-saving resource within a frame of dominant improvisation, and it was expected that through tourism the difficulties caused by a lack of capital and foreign currency, as well as market scarcity, could be solved.

In the 1960s, some countries in Central and South America began focusing on tourism as a means of developing their economies. Heritage, both tangible and intangible, was highlighted as an important resource upon which they could base their efforts. This chapter describes some of the types of heritage resources that exist in Latin America that have become an important element of the region's tourism product. It also examines some of the challenges facing leaders and heritage planners, the most troubling and difficult of which is political instability. Empirical cases from Peru, Guatemala, Nicaragua, and Argentina are presented to illustrate some of these important topics.

A heritage overview of Latin America

Based on the example of Spain, which was considered a valuable exemplar, and conscious that the prevalent form of mass tourism demanded the sun, the sea, and the sand (SSS), Latin American tourism development efforts were concentrated on the establishment of this beach-based product. Countries such as Bolivia, unable to offer the "sun and sand" option, concentrated on the heritage surrounding Lake Titicaca, on traces of a rich pre-colonial and colonial past, as well as a considerable jungle (Schlüter 2001). Countries such as Argentina chose to concentrate on their rich natural heritage, while Peru and a few other nations decided on cultural tourism, represented by the built heritage of the Inca Empire.

As years passed, tendencies changed and culture acquired greater importance in tourism, expressed by means of important products, revolving around the "monumental" heritage of aboriginal communities. Architectural sites increased in attractiveness, reinforced by non-material culture that provided a unique addition, such as clothing, gastronomy, festivities, and rites observed in public places and at fairs.

Latin America is endowed with many forms of heritage. Colonial architecture and historic cities abound and have become important destinations, and several have been placed on UNESCO's list of World Heritage Sites (Lew *et al.* 2008). Many of these cities face pressures of urban sprawl, increased population growth, and overuse. Brazil and Argentina are home to a variety of ethnic enclaves, such as Welsh and Germans, who immigrated from Europe early in the twentieth century and established European-looking communities with unique architecture, food, language, and attire (Ferguson 1995; Gade 1994). Indigenous living culture is an important part of the heritage product in Mexico, Guatemala, Honduras, Belize, Costa Rica, Panama, and the Andes Mountains of Peru and Bolivia (Cone 1995; Medina 2003; Mitchell and Eagles 2001; Mowforth *et al.* 2008). Likewise, striking ancient ruins and archeological sites in Mexico and Central America emanate an appeal like nowhere else on earth, and the cowboys of Argentina and the mega-festivals of Brazil attract hundreds of thousands of domestic, regional, and foreign tourists every year.

Cuisine is an important part of Mexico's heritage product and, like Mexico, Peru has begun using gastronomy to link its cultural heritage as a unique intangible heritage product to enhance the tourist experience. Many of Peru's gastronomical traditions were on the verge of being lost, so efforts have been made to rescue and preserve the traditional ingredients in ancient Andean cuisine while adopting modern preparation methods.

Even though Mexico was known for cultural heritage because of its Aztec sites, it was the Mayan culture that gave Mexico its best international visibility. Curiously, Cancun, an integrated tourism center on Mexico's Caribbean coast, created to satisfy the SSS demand of wealthier North American tourists, sparked the main interest in the region's cultural base. But the market segment to which it was geared did not develop, so that the mega-hotels were filled with charter tourists arriving mainly from Argentina and Brazil instead. However, the highway infrastructure created especially for Cancun's SSS tourism and the comfortable accommodations determined that it would be converted into a base for European tourists, who showed a marked interest in becoming familiarized with the cultures of Yucatan. Today, Cancun has been transformed into a service center for what is known as the "Maya World," an ambitious project that involves five countries over a span of 1,500 miles: Mexico, Guatemala, Belize, Honduras, and El Salvador. Work on this project is carried out with technical assistance from the European Union and various entities related to heritage management. Although Mexico is the most developed of these five countries and so far has benefited most from this effort, Guatemala is considered an important part of the project, as the Mayan heritage of Tikal is one of the most important in the region from an archeological point of view and exudes significant appeal for tourists.

South America also has pre-Columbian resources of great significance, such as the Incan culture best exhibited in the Machu Picchu fortress in Peru, discovered in 1911 by Hiram Bingham. It is near the city of Cuzco, itself an ancient Incan settlement of great importance on whose buildings the Spaniards erected their churches and other buildings. All of the area surrounding the fortress has salient cultural value, including the Sacred Valley of the Incas, and open fairs and markets where products grown and manufactured by local inhabitants are sold.

Like Cuzco, most Latin American cities, as already noted, have a rich colonial heritage. Many have been inscribed on UNESCO's World Heritage List (see Figure 12.1). For example, it is worth mentioning Antigua, the first capital of Guatemala, Cartagena de Indias (on Colombia's Caribbean coast), Ouro Preto (Brazil), and Quito and Lima, the capital cities of Ecuador and Peru respectively. When living cultural manifestations of the past are added to this rich architectural heritage, the attraction factor increases dramatically, as in the case of Salvador (Bahia) in northeastern Brazil, which is famous for its Carnival celebrations, although they are not as widely known and attended as the Carnival in Rio de Janeiro.

Perceived risk and tourism

Tourism is an extremely sensitive industry that rattles and dives with every threat of political, health-related, or natural disturbance. Strikes, violence, threats of war, or xenophobic policies may lead to severe downturns in arrivals at tourist destinations (Fernández Fuster 1974). Perceptions of risk occur when individuals face uncertainty about a trip or destination, which influences the purchase of a trip (Domínguez *et al.* 2001). According to the World Tourism Organization (UNWTO 1997), the way people react when faced with political instability is not always clear. Warlike situations that endanger many people's lives at a tourist destination may be considered less threatening for tourists than a single act of terrorism that affects relatively few people. Nevertheless, the threats occurring in remote destinations usually result in a greater impact (UNWTO 1997: 19).

High crime rates can also have negative effects on the success of tourism, and official government travel advisories play an important role in this. In the 1990s, Brazil was placed on the US government's travel warning list for the first time owing to problems in Rio de Janeiro. This resulting in a diminished demand for vacations to Brazil among Americans, because the country was not considered safe enough to visit (Schlüter 1993).

The media also play a similar role in the formation of risk images of a destination country. According to Sönmez and Graefe (1998), constant

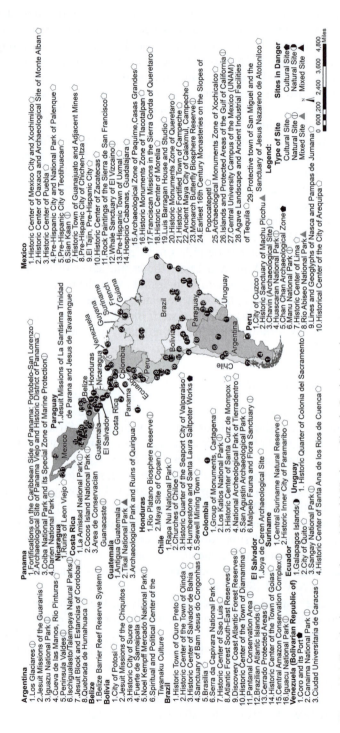

Figure 12.1 Developing countries and World Heritage Sites in Latin America

exposure in the media as a result of political or social instability causes a shift in demand to other, less dangerous places, despite cross-national relations and the potential appeal of the troubled destination. Despite these challenges, many developing countries that experience violence or other such crises try to attract tourists. In such cases, promotional efforts by official tourism organizations and agencies contrast starkly with the images of violence broadcast by the mass media, creating incongruity in the perceived image of the destination (Vettorazzi Mencos 1999: 233). Cohen (1993) refers to this incongruity in the context of Peru during the reign of a terrorist group, *Sendero Luminoso* (the Shining Path), which severely damaged the country toward the end of the twentieth century. With the image of a group of peasants, in colorful clothing, smiling, and welcoming visitors, with Machu Picchu in the background, news agencies spewed information about mass grave-sites, depicting soldiers and guerillas with dark masks and broadcasting situations of extreme poverty.

One of the most salient challenges facing tourism, and heritage tourism specifically, across much of Central and South America is political violence and war. Most countries in the region have experienced some form of conflict in small or large measure during the past forty years. Today, political instability and other forms of conflict affect tourism and heritage management in several countries (e.g., Guatemala, Mexico, Peru, and Colombia). Along with the efforts made as far back as the 1960s to grow tourism, political and social tensions started to arise, leading to internal conflicts of various magnitudes. These resulted in severe political instability throughout much of Latin America and included coups d'etat, rising guerrilla movements, and civil wars. Although these were more prominent in the 1960s after the triumph of the Cuban revolution, in many cases, they occurred far into the 1970s and 1980s. Even today, there are significant political tensions in Guatemala, Colombia, Ecuador, Peru, Venezuela, and Bolivia.

Leftist ideas prospered in many countries, and various groups willing to fight against the pre-established socio-political order arose. Political turbulence affected various forms of cultural tourism that were starting to develop. This was related not so much to what affected the heritage sites themselves but rather to the observable consequences on supply and demand (Teye 1986; Vukonić 1997). A considerable drop in demand occurred because of the perceived risk of visiting these places, but also on the supply side owing to a lack of investments, both local and from abroad, to develop the infrastructure necessary to satisfy the needs of tourists.

However, such situations are not always so clear in Latin America, as an example from Colombia demonstrates. Tourism officials tout Colombia as one of the most beautiful countries in South America, highlighting the attributes of monumental heritage at Cartagena de Indias, while the media concentrate on spreading word about the suffering of hostages taken by FARC (the Revolutionary Armed Forces of Colombia), a US- and EU-labeled terrorist group that has been perpetuating violence in the country for many years. Despite these tensions and present dangers, Colombia continues its counter-promotional efforts to attract tourists. In doing so, tourism officials emphasize the nation's present strong and stable economy and the fact that Cartagena de Indias hosted the seventh session of the World Tourism Organization's General Assembly, which was an apparent success. According to the Colombian National Tourist Office, the country is safe, except for a few areas. World Tourism Organization data suggest that the counter-promotional efforts have been a success, as Colombia received some 1,053,000 tourists in 2006, an increase of 18 percent over 2005, regardless of the unrest there.

Positive imagery may be useful for tourist offices to reposition their destinations once conflicts are over, which occurred in the state of Chiapas, Mexico, with the arousal of the Zapatistas. This led to a general lack of security and even specific targeting against tourists in Mexico City, which spread a sense of uncertainty throughout the whole country. Domínguez *et al.* (2001) indicated that, because of the 1994 indigenous rebellion, in Chiapas, where San Cristóbal de las Casas is considered to be one of the most attractive sites in the state because of its landscapes and culture, tourist arrivals decreased by nearly one third (31 percent). During the period 1994 to 1997, in which the Zapatista threats were strongest, tourism operators reported many trip cancellations to the conflict zone because of the climate of uncertainty.

Once the conflict subsided, the state Secretariat of Tourism Development made considerable efforts to recover and increase tourism using promotional resources that aimed to show the regained peace in the region and highlight its cultural attractions. The Secretariat's efforts had some success. In 1999, 1,478,353 tourists arrived in Chiapas, representing a considerable increase compared with 1994 arrivals. Tourists came mainly from Europe, motivated not only by ethnic, cultural, ecological, and adventure experiences but also by the notoriety of the Zapatista rebellion (Domínguez *et al.* 2001).

Nevertheless, despite this relative success, when a loss in tourism occurs as a result of long-term violence, recovery is slow, and recovery processes must include actions directed toward the supply side, which happened in Peru once the country experienced peace again following the capture in 1992 of the rebel leader of *Sendero Luminoso*.

An example of the dark days of Peru and beyond

As tourism began to grow in South America, Peru offered two very important attractions: Lima, the capital and national gateway with its rich colonial heritage; and the ancient fortress of Machu Picchu and its surroundings. Toward the end of the 1960s, tourists arrived sporadically at Cuzco, the gateway to Machu Picchu. Many of these travelers were American hippies with low travel budgets seeking to identify with a different culture. They used local means of transport—a significant improvement over the earlier days when the sites could only be reached by foot or donkey. Tello Rozas (2000) indicates that, in 1968, reacting to attention from UNESCO, the World Bank, and a group of tourism investors, the government prepared an inventory of the region's resources and thus became aware of the tremendous potential for the development of heritage tourism. The COPESCO Plan was created via an agreement between the Peruvian government and UNESCO. This plan facilitated additional support for tourism by creating appropriate legal frameworks and preparing and initiating a series of tourist routes.

The 1971–75 Development Plan set general guidelines, emphasizing the need to establish "self-support programs" based on agrarian reform and complemented with tourism and handicraft production. The next development plan for 1975–78 was launched, followed by the development plan for 1979–80. Both plans focused on the important role of heritage and tourism in regional development, together with mining and agriculture (Tello Rozas 2000).

In the early days, access to Cuzco was generally by plane. The trip could be extended to Puno, a population center located on Lake Titicaca, by means of an ancient and picturesque train ride that took several hours. On Titicaca, the floating islands made of "totora" roots by the Uros, an ancient people of the region, were visited and then the journey continued toward La Paz, Bolivia, or travelers could return to Lima.

Twenty-four kilometers from Puno is the Island of Tequile where tourism started in 1976 when backpackers arrived in Peru in large numbers, as they had throughout South America (Tello Rozas 2000). Agriculture, fishing, and the manufacture of textiles are the main activities of Tequile, with textile production being particularly remarkable because of its quality and originality. This resulted in Tequile and its textiles being included on the list of oral and intangible heritage of UNESCO in 2005. The sale of cloth allowed the Tequileans to obtain sufficient resources to buy boats with engines and start developing a tourism industry.

These attractions complemented others in the north of the country, including the Amazon Basin, which also became an important attraction in Peru. Thus, heritage tourism established a strong hold in Peru until the 1980s when the actions of *Sendero Luminoso* became more apparent. The Maoist group was considered one of the most dangerous and violent organizations in history with their objective being to replace Peruvian institutions with a peasant-based revolutionary regime and free Peru from foreign influences (Sönmez 1998). Their ideology was centered on weakening government authority and the Peruvian economy (Nuñez Salazar 2001). Their primary operations center was the city of Ayacucho, although they maintained a strong influence in Lima, a city that should, by all accounts, have continued to be an important tourist destination because of its rich architectural heritage. Nonetheless, the threat of violence and the campaigns associated with *Sendero Luminoso* transformed an important cultural meeting place into a dangerous city. This resulted in the historic center being nearly abandoned and avoided by tourists, and all new tourism services being relocated to the outskirts of town.

One feature of this group was its propensity to direct attacks against tourists. Among the most notorious acts of violence was that of June 1986, when an American tourist and six Japanese tourists were killed in an attack on the train between Cuzco and Machu Picchu. Another well-known and widely publicized incident occurred in 1990 when two French tourists were murdered while traveling by bus in the province of Andahuaylas (Nuñez Salazar 2001). According to Sönmez (1998), these attacks resulted in an immediate and acute decrease in tourism, from 350,000 international arrivals in 1989 to 33,000 in 1991. Between 1988 and 1993, many Western governments warned their citizens and potential business investors of the dangers of visiting and investing in Peru (Marsano Delgado 2000).

After the imprisonment of *Sendero Luminoso*'s leader in 1992, matters slowly began to change with a new-found growth in tourism after 1993. During the 1993–99 period, the annual average growth rate was 23 percent, while the accumulated rate for the same period was over 247 percent. As a consequence of the gradual recovery in tourist arrivals, the tourism balance of payments moved to a positive figure, foreign currency once again began to flow, and tourism businesses began to flourish (Marsano Delgado 2000: 50). Marsano Delgado (2000) adds that, from a poor and insufficient hotel infrastructure at the beginning of the 1990s, investments in infrastructure had a salient multiplier effect on the economy. Foreign investment between 1990 and 1999 rose to US$ 247 million and expanded at an annual growth rate of 17.3 percent.

The historical center of the city of Lima recovered and was incorporated into UNESCO's World Heritage List. In Cuzco and Machu Picchu, hotels were reconditioned and constructed. An elegant seventy-six-room hotel was opened with a view of the fortress of Machu Picchu, offering all the services required by a demanding clientele. Another hotel was inaugurated nearby following the guidelines of the now fashionable boutique hotels with room for seventy-six guests and oriented toward adventure- and heritage-based tourism. The train corridor between Cuzco and Puno was reconditioned, providing greater comfort and emphasizing quality food and meals on board.

Winds of change for Central America

Although Central America is a region known for natural beauty, diverse cultures, and remnants of ancient civilizations, it is also famous for hurricanes, earthquakes, volcanic activity, and political instability. The main issue affecting tourism arises, as it has in Peru and Colombia, from internal political conflict. In Central America, the legacy of instability is almost a way of life and has affected nearly all of its countries, including Nicaragua, Honduras, Guatemala, El Salvador, Panama and, to a lesser extent, Belize. These crisis conditions can be blamed, in part at least, for the region's lagging socio-economic development. These political factors created such a negative image of the region that, when Costa Rica sought to position itself in the 1980s as one of the world's leading eco-tourism destinations, the country used the argument that it was "different" from the rest of Central America, emphasizing the fact that its population gets along well, it has not seen war in over a century, and that the country does not even have a military force. This situation does, indeed, set Costa Rica aside from its neighbors.

Owing to its rich Mayan heritage, ideally expressed at Tikal, Guatemala has named itself the "Heart of the Maya World." Notwithstanding, as indicated by Vettorazzi Mencos (1999), the constant political–institutional conflicts and armed activities between the government and rebel guerrillas have notably affected the flow of tourists to the country because of a lack of security and the dangers created by a prolonged conflict of more than three decades. In several cases, tour buses in the Guatemalan highlands have been hijacked on their way to visit important heritage areas and indigenous communities. On a few occasions, individual tourists were singled out and murdered.

The perceived risk associated with Guatemala has had ups and downs. After 1986, when the military government failed and democracy was reinstated, tourism and other economic sectors grew by nearly 50 percent. The Nobel Prize awarded to Rigoberta Menchú helped to reinforce this image of change. Tourism continued to grow until 1993, albeit at lower rates, but new conflicts between 1994 and 1997 once again sent tourism plummeting. In the late 1990s, Guatemala's tourism grew at a much lower rate than in the other countries of Central America (Vettorazzi Mencos 1999).

A study conducted by Vettorazzi Mencos (1999: 242) indicated that the most negative perceptions about Guatemala from tourists' perspectives were lack of quality services and lack of security and freedom to travel within the country. The study also revealed that Guatemala lacks funds to be able to counteract through promotional efforts the country's negative security conditions. Other difficulties included a lack of product diversity, high service prices, inadequate airline connections, and concerns about safety and security.

The lack of resources has caused the government to appeal for assistance from international agencies, such as the Organization of American States (OAS) and the Inter American Development Bank (IDB), which prepared a series of development programs involving other countries. This help was geared toward the preparation of plans to boost education and create small and medium-sized enterprises that can be managed by the local community. With respect to heritage tourism, sites related to the Maya World and the colonial city of Antigua, Guatemala, were promoted. Another focus was oriented toward harmonious relationships between tourists and native populations.

The IDB project sought to create an ethnic tourism network to group the participating communities and companies together. The participants were to be trained through courses on hospitality, lodging, food service, guiding, handicrafts, and transportation. The groups interested in being involved in these tourist activities were expected to receive technical assistance to prepare business plans, establish quality standards, establish firms and associations, and request loans (http://forms.iadb.org/).

In turn, the Guatemalan government recognizes that tourism is a competitive sector with excellent opportunities, acknowledging it as a strategic sector because of its capacity to generate foreign exchange. Therefore, its objective is to adopt a long-term national policy that elevates tourism as a top priority in the national development strategy. The government also expects to promote a positive image in priority markets such as the United States, Europe, and Asia; promote tourism investments; facilitate free tourist transit within the region; continue their efforts to improve tourist security; and promote programs that will raise competitiveness, quality, and excellence in service provision (www.visitguat emala.com). Although not all of these proposals have succeeded, the positive consequences of what has been done are evident. At present, with 1.5 million international arrivals in 2006 (UNWTO 2007), Guatemala is now the second most visited country in Central America, preceded only slightly by Costa Rica.

Nicaragua is a different story. In tourism terms, it lags behind Guatemala, with 773,000 international arrivals in 2006. Nicaragua also has a history of political instability and a rich colonial heritage concentrated in the cities of Leon and Granada. Most of the colonial landscape is composed of churches built by the Spanish during their evangelizing efforts. During the government of General Somoza, who ruled the country with an iron fist and was considered an oppressive dictator by Nicaraguans and the international community, tourism began to grow, especially with increased arrivals from other Central American countries and the United States. Following his death, he was succeeded by members of his family until 1979, when an armed group called the *Frente Sandinista de Liberación* (Sandinista Liberation Front) entered Managua, the capital, and overthrew the government, taking power and thereby ending a "dynasty" that was established in 1934.

The first period of the Sandinista government was favorable toward tourism. Tourists arrived from different parts of the world, interested in becoming familiar with the revolution, the caudillo rule, and the rebel guerrillas. Hotels were developed, and many innovations took place, including an appraisal and promotion of the country's nature and cultural heritage.

Shortly afterward, Nicaragua experienced an economic crisis, sparked largely by the counter-revolution and a fiscal boycott by the United States in response to the new despotic Sandinista government. A war slowly began to unravel, which not only negatively affected the image of the country but also had direct consequences on the cultural heritage. Many historical documents were burned, and archeological pieces were destroyed or stolen. Cities such as Estelí and Leon were burnt or bombarded, and Granada's rich heritage was severely damaged. Toward the end of the 1980s, peace talks began, and the 1990 election was seen as a victory when a coalition party opposing the Sandinista movement was put into power.

Since 1990, Nicaragua has attempted to attract tourists and develop tourism. The country has a lot of potential, but so far it has failed to realize that potential. Nicaragua's cultural heritage does not have the monumental appeal of nearby Guatemala and, as for eco-tourism, it is difficult to compete with the success of Costa Rica. With respect to sun, sea, and sand tourism, Mexico and Belize are important competitors. In 1998, the National Tourism Office of Nicaragua was formed with the objective of achieving sustainable tourism development by involving the population through training programs established with the support of international organizations. As the image of internal security problems still persists, as in the case of Guatemala, a special department was implemented to enhance tourist safety. Another priority of the present government is to improve the quality of services in the hospitality sector.

Another important issue for Nicaragua and Guatemala, as well as the other Central American countries, is the development of community-based tourism centered on cultural heritage. Millán Vázquez *et al.* (2007) researched this potential among tourists from Spain. The study was based in Nicaragua and El Salvador and found that there is a high level of satisfaction associated with visiting heritage sites in the two countries. However, the study also identified a negative perspective—that the high cost of travel to the region does little to reach the tourism community and limits these types of travel experiences to wealthy tourists.

When human suffering and conflict become an attraction

The actions of the leftist rebels in nearly all Latin American countries were strongly opposed by their governments. In some cases, these struggles were so fierce that they brought about even more instability and high-risk situations. Regardless, through time, with considerable media exposure, these movements became tourist attractions and heritage foci.

For example, the Argentine Cuban Ernesto "Ché" Guevara, considered the most inspiring person of the leftist expansion in the region, died during an ambush in Bolivia in 1967 (Siles del Valle 2007), thereby becoming a mythical figure and hero. The Guevara phenomenon reached such an extent that T-shirts with his enigmatic smile based on a picture taken by Alberto Korda in 1960 and the revolutionary inscription "Till victory always" can be bought in far-flung places, such as markets in Kuala Lumpur, Malaysia, sparking a new kind of political expression (Felicetti 2007). Later, Guevara's remains were transferred to Cuba where he was honored with an imposing majestic funeral, but his open tomb, documents, and memories of his "passing through" remained in Bolivia. Considering the roots and background of his image, a group of villagers is developing the "Path of the Ché," a tourist route and tour package that highlights places where the revolutionary traveled and worked.

Another example of this leftist trend is Argentina, where repression was so stark that it earned the term "state terrorism" and brought about the intervention of human rights organizations from various parts of the world. The primary group at risk was young people in general, particularly university students, even though people were arrested on the street regardless of their nationalities or age. Many were taken to detention centers where, if lucky, they were released after a few days; others were tortured and killed without the government providing any information to their families about their fates. Between 1976 and 1982, 340 detention centers operated throughout the country. Even today, the number of missing persons still has not been determined. The National Commission for Missing Persons reported 8,960 missing people but, according to Amnesty International, the number is closer to 10,000–15,000. Some estimates even place the number at 30,000. According to the National Commission for Missing Persons (Comisión Nacional Sobre la Desaparición de Personas; CONADEP), approximately 90 percent of those who disappeared were murdered. Thirty percent of these were women, 10 percent of whom were pregnant at the time (Álvarez 2000).

When democracy was reinstated in 1983, the detention centers were deactivated, and those that belonged to military or police institutions were relocated. The main idea was to transform these sites into "memorial centers," but little progress has been made in that area. In Rosario, one of the most important cities in Argentina, the branch of the II Regional Unit was transferred to an ex-military plant. This unit was the center of the former mayor's office, and the police station operated during the last military dictatorship as a detention and torture center for hundreds of citizens from the province of Santa Fe. In 2005, the central area of the building was transformed into a "dry" square and opened to the public as a "space to honor the memory, the testimony, and the re-encounter of all."

A project also exists to transform the most macabre of all these detention centers, the Escuela de Mecánica de la Armada (Navy School of Mechanics; ESMA), in Buenos Aires. The ESMA ceased functioning as a school for navy officers after the coup d'etat of March 1976, when it was turned into a concentration camp. It is estimated that approximately 3,000–4,000 missing persons passed through this camp, of which only a few survived (Di Tella 1999).

Because these events are relatively recent and still remain fresh in the public memory, the transformation of these sites of atrocity into memorial centers and their management is not an easy task. As indicated by Ashworth and Hartmann (2005), the management of heritage sites that commemorate human tragedy and suffering faces many difficulties. One potential problem may occur when people react differently to the displays and interpretive setups that depict the cruelty enacted upon other human beings. In some instances, the visitor can even react unexpectedly by identifying more with the perpetrator than with the victim. In this particular case, it is also important to bear in mind that, even though no one condones the brutality suffered during the period of repression, the events that occurred are presently being shown exclusively from a one-sided perspective. No references are made to the conditions that led up to these dark events or to attacks by subversive groups.

Those who were abducted by the military and police force during the military dictatorship simply disappeared. Notwithstanding the imagined tragic ends of their children's lives, mothers at least wanted to know where they were, obtain more precise information as to what had occurred, and in the worst case to have a body to bury. As they were unable to obtain answers, fourteen women started meeting regularly at the Plaza de Mayo (May Square) in front of the Government House in Buenos Aires beginning in 1977 to petition for information about their missing family members (Bellucci 2000).

In the beginning, they were roughly "invited" to leave the site but, rather than giving up, they were soon joined by others and the movement grew and became organized. It became an organization called the "Mothers of Plaza de Mayo," which was later known around the world by the white kerchiefs they wore on their heads with the names of their missing children written on them. Each Thursday at 3:30 pm, a crowded time of day, they walked silently in circles in the midst of the square. Foreign tourists took time to observe them and in a way provided silent support.

The increasing importance of these silent marches was such that they became incorporated into the tourist guides of Argentina. At present, these "rounds" are no longer being carried out, but the search continues through other channels. Nevertheless, on the ground-tiles of the Plaza de Mayo, white kerchiefs remain painted in a circular form and have become a mandatory stop on the tourist circuits of Buenos Aires, as well as for those who individually care to visit the site.

Concluding remarks

The second half of the twentieth century found Latin American countries anxious to develop tourism as a development strategy, but the region was also faced with severe political and social crises that resulted in mass tourist aversions. These conflicts became something of a hallmark in Latin America and have become part of the current appeal. Violent acts were focused directly toward people and not at buildings or heritage sites. The damage suffered was mostly collateral, except in the case of Nicaragua, where violence increased to such an extent that a war started between the national forces and the opposition, also known as the "contras," which devastated the country, and from which it is still trying to recover.

Tourism was primarily affected by the perceived risk among tourists, who chose to visit safer places. The decreasing flow of tourist dollars created a lack of funds, which caused heritage conservation to suffer. Nonetheless, the recovery of tourism seems to have reacted quite quickly to the end of violence in Latin America. As the flow of tourists started to increase in post-conflict states, investments and the help of international organizations interested in heritage conservation were soon forthcoming. Although in some places such as Guatemala situations related to common crimes still persist as a result of social inequality, tourism is once again rising. This is in part a result of each country providing adequate resources and services for their markets.

The situation in Nicaragua is different because, prior to the worst periods of violence, the country had not been positioned as a significant tourist destination. Nicaragua also has unique heritage resources, many of which were destroyed in its battles, which was something that did not occur as widespread in other countries of the region.

Parallel to the recovery of traditional heritage tourism, a new form of tourism is arising based on the memory of atrocities perpetuated by oppressive government regimes. Although the number of people interested in this kind of "dark" tourism (or thanatourism) is increasing in the main markets, it remains to be seen whether there will be a national acceptance of this form of heritage. Because these tragic events are so relatively recent and therefore socially close, the population in general would tend to ignore these types of commemorations. Young people who did not go through these experiences are well aware of the situation through their history lessons and by oral tradition transmitted from their elders. Although they recognize what has happened, most of them have little desire to hold on to this element of the past. Most citizens of the troubled states of the region have a desire to move beyond this period, to live in the present, and to look forward to a future heritage that is devoid of such atrocities.

References

Álvarez, V. (2000) El encierro en los campos de concentración. In P. Lozano (ed.), *Historia de las Mujeres en la Argentina – Siglo XX*, pp. 67–90. Buenos Aires: Taurus.

Ashworth, G. and Hartmann, R. (2005) Introduction: managing atrocity for tourism. In G. Ashworth and R. Hartmann (eds), *Horror and Human Tragedy Revisited: The Management of Sites of Atrocities for Tourism*, pp. 1–14. New York: Cognizant.

Bellucci, M. (2000) El movimiento de las Madres de Plaza de Mayo. In P. Lozano (ed.), *Historia de las Mujeres en la Argentina – Siglo XX*, pp. 267–85. Buenos Aires: Taurus.

Cohen, E. (1993) The study of touristic images of native people: mitigating the stereotype. In D.G. Pearce and R.W. Butler (eds), *Tourism Research: Critiques and Challenges*, pp. 36–69. London: Routledge.

Cone, C.A. (1995) Crafting selves: the lives of two Mayan women. *Annals of Tourism Research*, 22: 314–27.

Di Tella, A. (1999) La vida privada en los campos de concentración. In F. Devoto and M. Madero (eds), *Historia de la Vida Privada en la Argentina – La Argentina entre Multitudes y Soledades. De los Años Treinta a la Actualidad*, pp. 79–106. Buenos Aires: Taurus.

Domínguez, P., Bernard, A. and Burguete, E. (2001) Imagen, seguridad y riesgo en destinos turísticos – el caso de San Cristóbal de las Casas, Chiapas, México. *Estudios y Perspectivas en Turismo*, 10(3/4): 251–67.

Felicetti, C. (2007) *Absolutamente Glam – Del Vestidito Negro a la Blusa Blanca, los 10 Magníficos del Vestuario Femenino*. Barcelona: Vergara.

Ferguson, J. (1995) Salsa and lederhosen. *Geographical*, 67(1): 22–24.

Fernández Fuster (1974) *Teoría y Técnica del Turismo*. Madrid: Editora Nacional.

Gade, D.W. (1994) Germanic towns in southern Brazil: ethnicity and change. *Focus*, 44(1): 1–6.

Getino, O. (1993) *Turismo y Desarrollo en América Latina*. México City: Ediciones Limusa.

Lew, A.A., Hall, C.M. and Timothy, D.J. (2008) *World Geography of Travel and Tourism: A Regional Approach*. Oxford: Butterworth Heinemann.

Marsano Delgado, J.M. (2000) Actividades económicas y financieras del turismo receptor del Perú durante la década de 1990. *Turismo y Patrimonio*, 1(2): 49–53.

Medina, L.K. (2003) Commoditizing culture: tourism and Maya identity. *Annals of Tourism Research*, 30: 353–68.

Millán Vázquez, F., López-Guzmán, T., and Caridad, J.M. (2007) Turismo comunitario en Centroamérica – Un análisis econométrico. *Papers de Turisme*, 41: 57–73.

Mitchell, R.E. and Eagles, P.F.J. (2001) An integrative approach to tourism: lessons from the Andes of Peru. *Journal of Sustainable Tourism*, 9(1): 4–28.

Mowforth, M., Charlton, C. and Munt, I. (2008) *Tourism and Responsibility: Perspectives from Latin America and the Caribbean*. London: Routledge.

Nuñez Salazar, C. (2001) Violencia y turismo – un ejemplo de Venezuela. *Estudios y Perspectivas en Turismo*, 10(3/4): 291–266.

Schlüter, R. (1993) Tourism and development in Latin America. *Annals of Tourism Research*, 20: 364–67.

—— (1998) Tourism development: a Latin American perspective. In W.F. Theobold (ed.), *Global Tourism*, pp. 216–30. Oxford: Butterworth Heinemann.

—— (2001) South America. In A. Lockwood and S. Medlik (eds), *Tourism and Hospitality in the 21st Century*, pp. 181–91. Oxford: Butterworth Heinemann.

Siles del Valle, J.I. (2007) *Los últimos días del Che – Que el sueño era tan grande*. Buenos Aires: Debate.

Sönmez, F. (1998) Tourism, terrorism and political instability. *Annals of Tourism Research*, 25: 416–56.

Sönmez, F. and Graefe, A. (1998) Influence of terrorism risk on foreign tourism decisions. *Annals of Tourism Research*, 25: 112–44.

Tello Rozas, S. (2000) Patrimonio, turismo y comunidad. *Turismo y Patrimonio*, 1(1): 151–63.

Teye, V. (1986) Liberation wars and tourism, development in Africa: The case of Zambia. *Annals of Tourism Research*, 13: 589–608.

Vettorazzi Mencos, R. (1999) La paz en Guatemala – Una oportunidad turística. *Estudios y Perspectivas en Turismo*, 8(2 and 4): 232–50.

Vukonić, B. (1997) *Tourism in the Whirlwind of War*. Zagreb: Golden Marketing.

World Tourism Organization (1997) *Seguridad en turismo – Medidas prácticas para los destinos*. Madrid: OMT.

—— (2007) *Tourism Highlights 2007 Edition*. Madrid: UNWTO.

13 Heritage tourism in Central and Eastern Europe

Duncan Light, Craig Young, and Mariusz Czepczyński

Introduction

Defining the region Central and Eastern Europe can be problematic. The region lacks clear geographical boundaries (Dingsdale 1999) and, while there is some consensus on its Western extent, there is less agreement regarding its Eastern borders. Consequently, a clear consensus on which countries are—and are not—included in the region is lacking. For the purposes of this chapter, we take Central and Eastern Europe to be composed of the following twenty states (see Figure 13.1): Albania; Belarus; Bosnia and Herzegovina; Bulgaria; Croatia; Czech Republic; Estonia; Hungary; Latvia; Lithuania; Macedonia (also known as the Former Yugoslav Republic of Macedonia); Moldova; Montenegro; Poland; Romania; the Russian Federation; Serbia; Slovakia; Slovenia; and Ukraine. This region also includes the Russian exclave of Kaliningrad. Overall, Central and Eastern Europe covers an area of over 19.2 million square kilometers (7.4 million square miles) and is home to 328.95 million people (see Table 13.1).

Within the region, a number of subregions are sometimes identified (see Dingsdale 1999). "Central Europe" is used to describe those countries (Czech Republic, Hungary, Poland, Slovakia, Slovenia) that historically and culturally have had the strongest ties with Western Europe (Croatia and Romania are also sometimes included in this group). The "Baltic States" is the term used to describe Estonia, Latvia, and Lithuania. Southeast Europe includes Albania, Bosnia and Herzegovina, Bulgaria, Macedonia, Montenegro, Romania, and Serbia, and sometimes Croatia. It also includes Kosovo, which declared independence from Serbia in January 2008 but has yet to be accepted universally as a sovereign state. The term "Balkans" is sometimes used interchangeably with "Southeast Europe" (although Croatia and Romania reject the label). "Eastern Europe" is sometimes used for the former Soviet Republics of Belarus, Moldova, and Ukraine (and we follow the UN definition in including the Russian Federation as part of Eastern Europe). Unless otherwise indicated, we use Central and Eastern Europe (CEE) in this chapter to describe the region as a whole.

Many of the states of CEE are relatively recent creations. Unlike Western Europe, much of the region was long under the control of various feuding

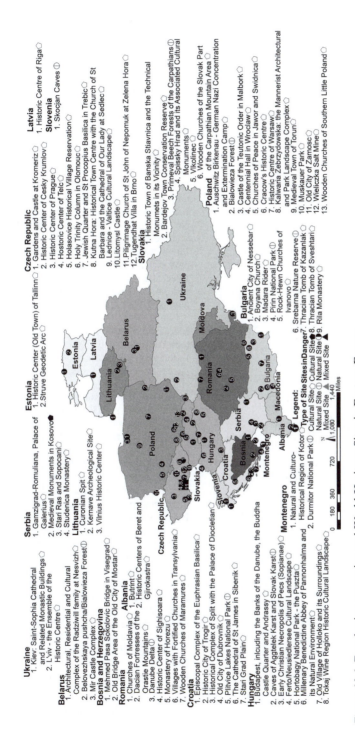

Ukraine
1. Kiev: Saint-Sophia Cathedral and Related Monastic Buildings ○
2. L'viv - the Ensemble of the Historic Centre ○

Belarus
1. Architectural, Residential and Cultural Complex of the Radziwill family at Nesvizh ○
2. Belovezhskaya pushcha/Bialowieza Forest ⊕
3. Mir Castle Complex ○

Bosnia and Herzegovina
1. Mehmed Pasa Sokolovic Bridge in Visegrad ○
2. Old Bridge Area of the Old City of Mostar ○

Romania
1. Churches of Moldavia ○
2. Dacian Fortresses of the Orastie Mountains ○
3. Danube Delta ⊕
4. Historic Center of Sighisoara ○
5. Monastery of Horezu ○
6. Villages with Fortified Churches in Transylvania ○
7. Wooden Churches of Maramures ○

Albania
1. Butrint ○
2. Historic Centers of Beret and Gjirokastra ○

Croatia
1. Episcopal Complex of the Euphrasian Basilica ○
2. Historic City of Trogir ○
3. Historical Complex of Split with the Palace of Diocletian ○
4. Old City of Dubrovnik ○
5. Plitvice Lakes National Park ⊕
6. The Cathedral of St James in Sibenik ○
7. Stari Grad Plain ○

Hungary
1. Budapest, inlcuding the Banks of the Danube, the Buda Castle Quarter and Andrassy ○
2. Caves of Aggtelek Karst and Slovak Karst ⊕
3. Early Christian Necropolis of Pecs (Sopianae) ○
4. Ferto/Neusiedlersee Cultural Landscape ○
5. Hortobagy National Park - the Puszta ○
6. Millenary Benedictine Abbey of Pannonhalma and its Natural Environment ○
7. Old Village of Holloko and its Surroundings ○
8. Tokaj Wine Region Historic Cultural Landscape ○

Serbia
1. Gamzigrad-Romuliana, Palace of Galerius ○
2. Medieval Monuments in Kosovo ●
3. Stari Ras and Sopocani ○
4. Studenica Monastery ○

Lithuania
1. Curonian Spit ○
2. Kernave Archeological Site ○
3. Vilnius Historic Center ○

Czech Republic
1. Gardens and Castle at Kromeriz ○
2. Historic Center of Cesky Krumlov ○
3. Historic Center of Prague ○
4. Historic Center of Telc ○
5. Holasovice Historical Village Reservation ○
6. Holy Trinity Column in Olomouc ○
7. Jewish Quarter and St Procopius Basilica in Trebic ○
8. Kutna Hora: Historical Town Centre with the Church of St Barbara and the Cathedral of Our Lady at Sedlec ○
9. Lednice - Valtice Cultural Landscape ○
10. Litomysl Castle ○
11. Pilgrimage Church of St John of Nepomuk at Zelena Hora ○
12. Tugendhat Villa in Brno ○

Slovakia
1. Historic Town of Banska Stiavnica and the Technical Monuments in its Vicinity ○
2. Bardejov Town Conservation Reserve ○
3. Primeval Beech Forests of the Carpathians ⊕
4. Spissky Hrad and its Associated Cultural Monuments ○
5. Vlkolinec ○
6. Wooden Churches of the Slovak Part of the Carpathian Mountain Area ○

Poland
1. Auschwitz Birkenau - German Nazi Concentration and Extermination Camp ○
2. Bialowieza Forest ⊕
3. Castle of the Teutonic Order in Malbork ○
4. Centennial Hall in Wroclaw ○
5. Churches of Peace in Jawor and Swidnica ○
6. Cracow's Historic Centre ○
7. Historic Centre of Warsaw ○
8. Kalwaria Zebrzydowska: the Mannerist Architectural and Park Landscape Complex ○
9. Medieval Town of Torun ○
10. Muskauer Park ○
11. Old City of Zamosc ○
12. Wieliczka Salt Mine ○
13. Wooden Churches of Southern Little Poland ○

Estonia
1. Historic Center (Old Town) of Tallinn ○
2. Struve Geodetic Arc ●

Bulgaria
1. Ancient City of Nessebar ○
2. Boyana Church ○
3. Madara Rider ⊕
4. Pirin National Park ⊕
5. Rock-Hewn Churches of Ivanovo ○
6. Srebarna Nature Reserve ⊕
7. Thracian Tomb of Kazanlak ○
8. Thracian Tomb of Sveshtari ○
9. Rila Monastery ▲

Montenegro
1. Natural and Culturo-Historical Region of Kotor ○
2. Durmitor National Park ⊕

Latvia
1. Historic Centre of Riga ○

Slovenia
1. Skocjan Caves ⊕

Legend:
Type of Site — Sites in Danger
Cultural Site ○   Cultural Site ●
Natural Site ⊕   Natural Site ●
Mixed Site ▲   Mixed Site ▲

0 180 360 720 1,080 1,440
Miles

Figure 13.1 Developing countries and World Heritage Sites in Eastern Europe

Table 13.1 Basic data on the countries of Central and Eastern Europe

Name	Area (sq km)*	Population (millions)*	Notes†
Albania	28,748	3.5	Independent state since 1912. Potential future member of EU
Belarus	207,595	9.8	Independent state since 1991 (formerly part of USSR). Has shown no interest in EU membership
Bosnia and Herzegovina	51,129	3.8	Independent state since 1992 (formerly part of Yugoslavia). Potential future member of EU
Bulgaria	111,000	7.7	Independent state since 1878. EU member since January 2007
Croatia	57,000	4.4	Independent state since 1992 (formerly part of Yugoslavia). Applied for EU membership in February 2003
Czech Republic	79,000	10.3	Independent state since 1993 (formerly part of Czechoslovakia). EU member since May 2004
Estonia	45,000	1.3	Independent state since 1991 (formerly part of USSR). EU member since May 2004
Hungary	93,000	10.1	Independent state since 1920. EU member since May 2004
Latvia	65,000	2.3	Independent state since 1991 (formerly part of USSR). EU member since May 2004
Lithuania	65,300	3.4	Independent state since 1990 (formerly part of USSR). EU member since May 2004
Macedonia	25,713	2.05	Independent state since 1992 (formerly part of Yugoslavia). Applied for EU membership in March 2004
Moldova	33,700	3.3	Independent state since 1991 (formerly part of USSR). Interested in EU membership but no application yet
Montenegro	13,812	0.6	Independent state since 2006 (formerly part of Yugoslavia and subsequently Serbia and Montenegro). Potential future EU member
Poland	313,000	38.1	Independent state since 1919 (although borders substantially modified after the Second World War). EU member since May 2004
Romania	238,000	21.6	Independent state since 1878 (borders substantially modified after the First World War). EU member since January 2007
Russian Federation	17,075,400	142.1	Independent state since 1991. Has not shown interest in EU membership

Table continued next page

Table 13.1 (continued)

Name	Area (sq km)*	Population (millions)*	Notes†
Serbia‡	51,129	10.1	Independent state since 2006 (formerly part of Yugoslavia and subsequently Serbia and Montenegro). Potential future EU member
Slovakia	49,000	5.4	Independent state since 1993 (formerly part of Czechoslovakia). EU member since May 2004
Slovenia	20,000	2.0	Independent state since 1992 (formerly part of Yugoslavia). EU member since May 2004
Ukraine	603,700	47.1	Independent state since 1991 (formerly part of USSR). Interested in EU membership but no application yet

*Source: http://europa.eu/abc/european_countries/index_en.htm
†References to independence refer to the state in its current form.
‡Including Kosovo. At the time of writing, Kosovo's independence had not been fully recognized by the world community.

empires. For example, much of Central Europe was under the influence of the Austro-Hungarian and Prussian empires, while the lands to the south and east fell under the influence of the Russian and Ottoman empires. The ebb and flow of these imperial powers have left diverse (and frequently contested) legacies. From the eighteenth century onward, nationalist movements arose throughout the region to campaign for independence and self-determination. Despite these efforts, independent states were late to form in the region. A few states (such as Bulgaria and Romania) attained independence during the nineteenth century. Others (such as Czechoslovakia and Poland) achieved independence after the First World War. The collapse of communist regimes between 1989 and 1991 heralded a new era of political volatility and a radical redrawing of the political map of Europe. Thus, the Soviet Union fragmented into a number of independent states in 1991. Czechoslovakia peacefully split in 1993 to form the Czech Republic and Slovakia. Finally, in the conflicts of the early 1990s, Yugoslavia fractured into six independent states (the last of which, Montenegro, declared its independence as late as 2006). In total, fifteen of the twenty states of Central and Eastern Europe have attained independence within the past two decades.

At first sight, Central and Eastern Europe may not appear to be a part of the developing world. Many of these countries enjoy a high level of prosperity and, indeed, ten of them are now members of the European Union (EU). United Nations figures appear to confirm this premise, in that twelve of the CEE countries are in the "high" category of Human Development Index scores, with the remainder (mostly the eastern states) being in the "medium" category (UNDP 2006). But the situation is not quite so straightforward.

While all the Western European states are classified by the World Bank as "high income," only five of those in Central/Eastern Europe (Czech Republic, Estonia, Hungary, Slovakia, and Slovenia) are in the same category. The remaining fifteen CEE states are classed as "middle income" (World Bank 2008), a characteristic they share with many other states in the developing world. Although most of the states of CEE are not as affluent as those of Western Europe, they initially appear to have more in common with the latter group than they do with the countries of the developing world.

On the other hand, what definitively sets CEE apart from Western Europe is the region's experience during the twentieth century of a political–economic system known as state socialism or, more commonly, communism. In 1922, the Union of Soviet Socialist Republics (USSR) was established, which included parts of modern-day Belarus and Ukraine, as the world's first state founded on the principles of Marxism–Leninism. During the Second World War, the USSR annexed Estonia, Latvia, Lithuania, and Moldova, and, at the end of the war, Soviet armies occupied Bulgaria, Czechoslovakia, Hungary, Poland, Romania, and the eastern part of Germany. Socialist regimes loyal to the Soviet Union were installed in all of these countries, while socialists also took control independently in Albania and Yugoslavia. Socialist regimes attempted to eradicate the market economy. Instead, they operated on the basis of state ownership of the means of production and central planning of the economy by the state with the aim of eliminating social and geographical inequalities (Young 2005a). Central planning was accompanied by a one-party political system in which the Communist Party assumed total political, economic, and social control. As a result, the various states of Central and Eastern Europe and the Soviet Union experienced a model of political and economic development that was radically different to that of Western Europe. Indeed, the term "Eastern Europe" was widely used in the West as a synonym for the socialist countries of Europe.

The socialist regimes in Central and Eastern Europe collapsed between 1989 and 1991, while the Soviet Union dissolved in 1991. These countries entered the "post-socialist" period, which has been characterized by dramatic experiences of change and "transition." With a few exceptions (such as Belarus), the CEE countries have sought (although in different ways and at different paces) to dismantle the political, economic, and social structures of socialism and replace them with the capitalist and democratic models of Western Europe. In the process, these countries have undergone radical neo-liberal economic restructuring and massive social transformations. Arguably, it is these shared experiences (and legacies) of state socialism, along with the common experiences of post-socialist restructuring, that mark out the distinctiveness of CEE as a region (cf. Stenning 2005) and, at the same time, differentiate it from Western Europe (this is also the reason why Greece, which has not experienced state socialism, is not included here as part of CEE).

In addition, this recent experience of state socialism and subsequent post-socialist restructuring is also shared by many states in the developing world

(Herrschel 2006). Moreover, much of the developing world (whether formerly socialist or not) has shared with CEE recent experiences of neo-liberal transformations and "structural adjustment." As such, there are further similarities between much of CEE and many states in the developing world. For example, in some CEE countries, processes of state-building are much weaker than in Western Europe (and sometimes considerably less strong than EU accession might suggest). In addition, there are many instances where other (at times local) economic and political actors (often operating at the fringes of legality) are important in governance structures. Like much of the developing world, some CEE states are typified by economies that would not completely equate with Western European notions of functioning market economies. Various individual and household survival strategies have been developed to cope with the increased social polarization that post-socialist marketization has brought about. These include non-market, illegal, and semi-legal practices, including subsistence production; barter and exchange; and the existence of "mafia"-type organizations. In addition, the period of Soviet domination in CEE during the state-socialist period can be regarded as a form of colonialism in "Eastern Europe." In this sense, there are similarities between the experiences of post-socialism in CEE and of post-colonialism in many parts of the developing world. Both situations involve the emergence (or creation) of new nation-states from old empires. In addition, there is a process of redefining senses of personal and national identity in which the creation and use of heritage resources play a key role. Overall, then, CEE as a region presents a distinctive set of issues and problems that are sufficiently different from Western Europe to merit analysis in their own right and, for this reason, the inclusion of the region in a collection of essays about heritage tourism in the developing regions is fully justified.

This chapter examines contemporary heritage tourism in Central and Eastern Europe. For reasons of space, it focuses principally on the region's built heritage. Like Western Europe, the region's complex and turbulent past has left a diverse range of resources for heritage and tourism. Many of these places are of international significance, and CEE is home to 115 World Heritage Sites (UNESCO 2009), some 13.1 percent of the global total. The discussion begins by considering the recent past—the socialist era—to provide an essential context for understanding the current (post-socialist) situation. The chapter then outlines the key types of heritage tourism emerging under post-socialism. One theme that underpins this discussion is the ideological and political role of heritage tourism in CEE. In particular, throughout this region, heritage plays a key role in the construction and projection of national identities for both "internal" and "external" audiences (Light 2001).

Heritage tourism in the socialist era

Tourism, like every other aspect of social life, had important ideological connotations for socialist regimes and was interwoven with educational and political projects (Hall 1991). Socialist states actively supported domestic tourism by providing holiday opportunities (of an organized collective nature) for the working classes. During the 1960s, the citizens of socialist states had increased opportunities for international travel, albeit confined to other countries within the socialist "bloc." From the 1960s onward, socialist states also opened to Western tourists, partly from a desire to generate hard currency but also to demonstrate the achievements of socialism to Westerners. The organization of tourism in socialist states was mainly controlled by state travel offices and workers' travel organizations, although in some countries (particularly Poland, Hungary, and Yugoslavia) private sector tourism flourished, especially in the most desirable mountain and coastal resorts.

To understand the importance of heritage tourism in socialist regimes, it is necessary to consider socialist conceptions of time and history. Socialism was a political project based on a radical break with the past and with its eyes firmly fixed on the future. Its aim was no less than the creation of a new society and a "new man" (Boia 1999). In such circumstances, socialist regimes might have been expected to be indifferent or even hostile to the past and its various legacies. However, the situation was more complex. All political regimes seek to legitimate themselves through anchoring or grounding their existence in the past. They seek to create a narrative of the "national past" that will become the focus of popular allegiance, thereby conferring legitimacy upon the regime. This is achieved in various ways including commemoration and the creation of ceremonies and traditions. Lacking the popular support of a democratic mandate, socialist regimes had to seek legitimacy in other ways. One of these was through presenting socialism as the inevitable pinnacle of the country's historical trajectory. Thus, socialist regimes energetically rewrote history to create official narratives that were often starkly at odds with popular memory.

Given the importance of history to socialist regimes, it is unsurprising that they used heritage tourism as a source of legitimation. Moreover, heritage was an important element of regional economies and ideological propaganda. Thus, the state was a major actor in the provision of heritage tourism through organized activities such as school trips (often compulsory as part of the educational curriculum), subsidized excursions, and organized visits to factories or other socialist enterprises. In most socialist regimes, this support for heritage tourism was part of a broader policy of promoting domestic tourism. Such tourism was viewed as a means of developing social integration, patriotism, and support for the communist ideal and, indeed, the remaining socialist states in the world continue to use domestic tourism in this way (Kim *et al.* 2007).

Among the most important resources for heritage tourism in socialist regimes were the memorials, shrines, and museums built to commemorate the revolutionary traditions of the communist movement itself. Socialist regimes were eager to commemorate their leaders, a development initiated in the Soviet Union with the construction of Lenin's mausoleum in Moscow. Other countries followed the Soviet example. For example, in Poland, there were museums dedicated to Lenin in Warsaw, Kraków, and the small village of Poronin. Local communist leaders and activists were also commemorated. Among the many examples were a museum dedicated to Rosa Luxemburg in Leipzig (East Germany), Georgi Dimitrof's mausoleum in Sofia (Bulgaria), and Nicolae Ceauşescu's birthplace in Romania. At Kumrovec, Yugoslavia, an open-air museum was constructed around the birthplace of Josip Broz Tito. Almost every major city in CEE had a museum or shrine dedicated to the revolutionary and/or workers' movement, and these were widely promoted in tourist guides. In addition, a whole series of lesser heroes and activists were commemorated at the local level. Monuments and ideological places of worship became particular places of semi-religious "pilgrimages" by the masses, although attendance was not always voluntary. Furthermore, as figures such as Marx and Lenin (as well as other local leaders) figured prominently in school curricula, there were also many educational visits to such places. A new "cult" accompanied the construction of the "socialist heroes" or "saints" in every socialist country of the region (see Rembowska 1989; Satjukow and Gries 2002). Very similar practices are also apparent in various states in East Asia that continue to adhere to the doctrines of state socialism (Henderson 2007).

Alongside the celebration of the heritage of revolutionary socialism, many of the CEE states were also eager to commemorate their struggle against fascism and Nazi Germany. Indeed, the heroic opposition to fascism and Nazism became a foundation myth of many socialist states (see Koonz 1994). Thus, new memorials and commemorative places were created to remember the victims of Nazi atrocities, although the many war crimes committed by the Red Army did not receive similar attention. In the Soviet Union, school children were taken to visit sites of heroic battles such as Stalingrad (Volgograd), Kursk, and Lenino, while huge museums of the "Great Patriotic War" were established in Moscow, Kiev, and Minsk. Former Nazi concentration camps or places of mass murder—such as Auschwitz in Poland, Oranienburg in East Germany, or Lidice in Czechoslovakia—became sites for visits by organized groups and school parties. At such places, socialist states commemorated the victims of fascism, but they were also places to assert the ideological agenda of socialism. By equating fascism with capitalism, and the West in general, the defeat of fascism was presented as one step toward the eventual global triumph of socialism (Koonz 1994).

Socialist regimes also created their own, new heritages through celebrating their own achievements. New towns, public buildings, factories and industrial complexes, power stations, roads, and railways were interpreted and promoted as the pride of the new system that socialism was creating. Examples include the steelworks and associated town of Nowa Huta, Poland (Dawson 1999), the "Iron Gates" dam, which linked Romania and Yugoslavia, and the Lenin Shipyard in Gdańsk, Poland. Many countries followed the Soviet Union in constructing monumental public buildings, and examples of Stalinist "wedding cake" architecture can be found in Riga, Prague, Warsaw, and Bucharest.

Perhaps the most famous example is the Red Army Theatre in Moscow, the overall plan of which is in the shape of a five-pointed star (Cohen 1995). In other instances, monumental public spaces were established to celebrate socialism. One of the best examples is *Ploshchad Svobody* (Freedom Square) in Kharkiv, Ukraine, which is designed in the shape of a hammer and sickle (Czepczyński 2008). Such monuments of socialism were extensively promoted in tourist guides of this era. In the same way, the new heritage of the working classes also features prominently in such guides. For example, one Polish guide to Budapest from the 1970s (Olszański 1970) dedicates twelve pages to the industrial working class district of Csepel, while allocating only seven pages to the city's museums. Similarly, such places were promoted to international tourists as a way of demonstrating the ideological superiority of state socialism.

Socialist regimes also faced the problem of reinterpreting and repositioning heritage resources from the pre-socialist era, many of which were distinctly unwelcome heritages. Many of the old architectural and artistic heritages were clearly connected with the former socio-economic systems (such as capitalism and feudalism) and therefore were discordant with the values and agenda of socialism. Socialist regimes adopted various strategies toward such heritages. In some cases, the intention was to eliminate entirely from the landscape those buildings that were associated with the former regime. Thus, some historic buildings were deliberately destroyed, such as the Church of Christ the Saviour in Moscow, which was demolished in 1931 on Stalin's orders (Sidorov 2000). Similarly, numerous historic buildings and churches were destroyed in the remodeling of Bucharest in the 1980s (Cavalcanti 1997). Others (such as churches or palaces associated with the former aristocracy) were simply left derelict, and some, even those of important historical and architectural value, did not survive until the end of the socialist era. Even monuments that were significant places for national identity were sometimes neglected, such as Trajan's Bridge in Romania (Light and Dumbrăveanu-Andone 1997). In other cases, royal, imperial, aristocratic, or religious buildings were reconfigured and repositioned to give them entirely new meanings that were appropriate to the age of state socialism. Thus, Bran Castle in Romania (a former royal palace) was taken into state ownership and reopened as a museum of feudal history and art (Prahoveanu and Coşuleţ 1985). Similarly, many churches in the Soviet Union were used as warehouses, while the most prominent (such as Saint Isaac's Cathedral in Leningrad) were turned into museums of religion and/or atheism. Others were simply abandoned and neglected.

But in other cases, socialist regimes made considerable investments in the preservation and upkeep of historic environments. One of the best examples was the reconstruction of Warsaw Old Town in Poland that had been largely destroyed by the Nazis in 1944. The Old Town was rebuilt according to the eighteenth-century paintings of Canaletto, while all nineteenth-century changes were ignored as "ideologically malignant." So effective was the reconstruction that Warsaw's Old Town was added to the World Heritage List in 1980. Similar rules were implemented in rebuilding other Polish cities such as

Gdańsk and Wrocław, where any traces of "German tradition" were excluded. Indeed, the "Polish school of conservation" became renowned for its thorough state-sponsored historic reconstructions (Ashworth and Tunbridge 1999). Similarly, a number of socialist regimes created open-air and ethnographic museums that preserved and interpreted rural heritages and traditions. Such museums were a way of preserving historical and regional building styles, but they were also a way of "museumifying" the pre-socialist past and thereby highlighting the new future that socialism was constructing.

From the 1970s onward, some socialist regimes sought popular legitimacy by appealing to national sentiments and adopted a position of being "socialist in substance and national in form." Consequently, these regimes sought to present themselves as guardians of the national heritage, and there was greater attention to maintaining historic places and buildings, albeit in a selective way that was appropriate to political needs (Nawratek 2005). In this context, historic cities such as Leningrad (Soviet Union), Kraków (Poland), Prague (Czechoslovakia), and Gjirokastër (Albania) underwent extensive restoration that proved popular with Western heritage tourists. The East German government started to embrace parts of the Prussian heritage to promote national pride and, consequently, it set about reconstructing some of Berlin's major churches and Prussian-era buildings. Indeed, the regime even set about creating a heritage quarter—Nikolaiviertel—in East Berlin in an attempt to connect the capital of the socialist state with an earlier medieval past. It soon proved popular with heritage tourists, both domestic and international.

While socialist states had a mixed record with respect to their historic and cultural heritage, their treatment of the region's natural environment and heritage was much bleaker (see Carter and Turnock 1996). Despite commitments to environmental protection, socialist regimes tended to disregard environmental concerns in the drive for industrialization. The most extensive and visible impact was in the region of sulfurous brown coal extraction in the southern part of East Germany, northwestern Czechoslovakia, and southwestern Poland. In addition to its impacts on human health, atmospheric pollution also caused extensive defoliation and damage to forests in the region. Moreover, the cultural heritage was equally threatened in historic cities such as Kraków where air-borne pollutants (largely from the socialist-era steelworks at nearby Nowa Huta) caused serious damage to sandstone sculptures and buildings. Rural environments fared little better in the drive for agricultural intensification. Rural landscapes were reclaimed for agricultural use, including unique and fragile ecosystems such as the Danube Delta in Romania. The excessive use of fertilizers and pesticides led to soil pollution and erosion and, in some cases, deterioration in water quality. Rural traditions and heritage were further eroded by rural depopulation and the drive to create an urban, industrial, working class. The most notorious example was the (short-lived) 1980s "systematization" program in Romania, in which the country's president sought the demolition of thousands of villages and the forced movement of people to new agro-industrial towns.

In summary, during the socialist era, CEE had extensive resources for a wide variety of forms of cultural and heritage tourism. In many cases, such tourism was actively encouraged by the state, particularly given the importance of history as a source of legitimation for socialist regimes. However, although the region was never actually closed to Western tourists, the level of demand from non-socialist countries was low. The result was that the principal source of demand for heritage tourism was the domestic market, and the region's abundant resources for heritage and cultural tourism were largely unknown in the West.

Heritage tourism in the post-socialist era

Since the fall of the socialist regimes, international tourism in many of the CEE countries has experienced a dramatic growth. In the immediate post-socialist period, the most dramatic growth was in those countries (particularly the former East Germany, Czechoslovakia, and Hungary, and later Poland) that bordered the European Union (Hall 1995, 1998). From the late 1990s onward, a number of countries further east (including the Baltic States, Slovenia, and Bulgaria) have also experienced increasing international tourism. Croatia recovered swiftly from the conflicts of the early 1990s and re-established itself as a destination for coastal holidays. However, as Table 13.2 shows, some parts of CEE have lagged behind in terms of international tourist visits. In

Table 13.2 International tourist arrivals in the countries of Central and Eastern Europe

Country	International tourist arrivals 2005 (millions)
Albania	0.046
Belarus	0. 91
Bosnia and Herzegovina	0.21
Bulgaria	4.84
Croatia	8.47
Czech Republic	6.34
Estonia	1.90
Hungary	10.05
Latvia	1.12
Lithuania	1.80*
Macedonia	0.20
Moldova	0.023
Poland	15.20
Russian Federation	19.94
Romania	1.43
Serbia and Montenegro	0.73
Slovakia	1.52
Slovenia	1.56
Ukraine	15.63*

*2004 data.
Source: World Tourism Organization (2006).

Southeast Europe (Albania, Bosnia and Herzegovina, Macedonia, Montenegro, and Serbia), international tourism was significantly depressed by conflicts in the former Yugoslavia, and these countries are only slowly repositioning themselves within the European tourist market. Similarly, international tourism demand is low in many parts of Eastern Europe. The former Soviet Republics of Belarus and Moldova continue to be associated with political uncertainty and, while international arrivals are higher in the Russian Federation and Ukraine, a significant component of this traffic is likely to be derived from neighboring countries and cross-border trading.

Much of the early post-socialist boom in tourism in CEE was associated with heritage tourism as many visitors from Western Europe were eager to visit the historic capitals of Central and Eastern Europe (Bratislava, Budapest, Prague, Warsaw, and later St. Petersburg and Ljubljana). These cities remain extremely popular but, over time, Western cultural tourists have also turned their attention to countries further east. Thus, historic cities such as Tallinn, Riga, Sofia, and increasingly Bucharest have also established themselves as destinations for heritage tourism. The development of (heritage) tourism within CEE has been significantly aided by the growth of low-cost airlines that have made many more cities accessible destinations. Thus, a number of secondary historic cities and regional capitals (such as Brno, Debrecen, Gdańsk, Krakow, Plovdiv, Lviv and, most recently, Bucharest) have also established themselves as short break and weekend destinations. Once political stability returned to Croatia, the country's historic cities (e.g., Split and Dubrovnik) quickly re-established themselves as centers for cultural tourism.

However, heritage tourism in CEE is not only focused on urban areas. The region has established itself as a major destination for rural tourism (Hall 2004). In many countries, rural lifestyles and traditions survive, the equivalents of which have almost entirely disappeared from Western Europe. Such a heritage, including strong traditions of village life and sustainable non-intensive agricultural practices, has survived almost unscathed socialist-era efforts to collectivize and intensify agriculture. This rural heritage has proved very attractive to Western tourists seeking an experience of the pre-modern rural "Other." At the same time, the development of "agro-tourism" (farm-based stays) is a welcome boost for the rural economy while diversifying the opportunities for rural communities. Countries such as Bulgaria, Latvia, Lithuania, Poland, Romania, and Slovakia have enthusiastically promoted rural tourism, and the activity receives EU support. Recognizing their rural heritage as a unique selling point, images of the pastoral and rural feature prominently in the "official" tourism promotional materials produced by the CEE countries (Light 2006; Morgan and Pritchard 1998).

In addition to these forms of heritage tourism, CEE also contains other, unique heritage resources that set it apart from other areas of the developing, or recently developed, world. First, the heritage of the region's Jewish population and the Holocaust is an increasingly prominent feature of tourism, particularly in countries such as Germany and Poland. Since the collapse of socialist regimes, there has been a significant increase in the numbers of tourists visiting such sites, and there has been a shift in how they are interpreted. In particular, socialist interpretations that depicted the socialists as liberators are no longer dominant, and there is now more emphasis on commemorating the victims of the Holocaust. New museums and memorials have appeared, for example the Jewish Museum and the Holocaust Memorial in Berlin (see Jansen 2005). In other cities, former "Jewish quarters" (such as Josefov in Prague in the Czech Republic) have been restored and promoted to heritage tourists. Elsewhere, film-induced tourism has led to the commodification of Jewish heritage. Perhaps the best example is the Kazimierz district of Krakow, which has witnessed the preservation of its Jewish heritage in association with "*Schindler's List* tourism" (Murzyn 2006).

Another distinguishing feature of the heritage tourism product of CEE is the legacy of the socialist past. This phenomenon has been term termed "socialist heritage" or "Communist heritage tourism" (Frank 2006; Light 2000a, 2000b; Young and Light 2006). It involves visits to places associated with the socialist past or to sites that interpret or commemorate that past. Such tourism takes a variety of forms. For example, in some CEE capital cities, there is now a "Museum of Communism" (such as the DDR Museum in Berlin or the "Museum of Communism" in Prague). In other cases, the city's national museum includes a section dedicated to the socialist era (such as the gallery of twentieth-century history in the National Museum of Hungary in Budapest). Although such museums often claim to give a picture of everyday life under socialism, in many instances the focus is only on the negative aspects of socialism or "Soviet occupation" (such as the Museum of Occupation 1940–91 in Riga). In Berlin, the "Stasi Museum," Checkpoint Charlie museum, and the Berlin Wall interpretive center focus on the oppressive nature of the socialist state. In other instances, socialist-era statues have been removed from their original positions and placed in open-air museums, notably in *Szoborpark* (Statue Park), Budapest, Grūto Parkas (Grūtas Park), Lithuania, and Kozłówka in Poland. The few surviving sections of the Berlin Wall have become a major tourist attraction in the city (Frank 2006; Light 2000a; Timothy 2001). In other contexts, the heritage focus is more on commemorating histories of anti-Communist resistance, particularly in cities associated with the fall of state socialism such as Gdańsk, Nowa Huta, and Leipzig. In Poland, Gdańsk and Nova Huta incorporate key events and figures associated with the 1980s Solidarity movement in their contemporary identities. Gdańsk also hosts the *Drogi do Wolności* (Roads to Freedom) exhibit about the Solidarity movement. In Moscow, socialist-era statues are still displayed in the Park of Arts (*Park Isskustv*) and Victory Park (*Park Pobedy*), and link in complex ways to post-Soviet identities in Russia (Forest and Johnson 2002; Forest *et al.* 2004).

Key public buildings from the socialist era are also now utilized as heritage tourism resources, although the ways in which the histories of the buildings are represented varies considerably and does not always reflect their socialist origins. Examples include the vast "Seven Sisters" tower blocks in Moscow;

Warsaw's Palace of Science and Culture (see Cohen 1995; Dawson 1999; Young 2005a); Bucharest's "House of the People" (Light 2001); and the monumental triumphant buildings of central Minsk and Kiev. The architectural and town planning heritages of state socialism are also entering the tourism product. There are interpretive boards along the former Stalinallee/ Karl Marx Allee in the former East Berlin, and in the Poruba distract of Ostrava, Czech Republic. Industrial towns built by state-socialist regimes are another example. Most notable here is the town of Nowa Huta in southern Poland, built in the 1950s to house the workers at the Lenin Steelworks, which is now promoted as a part of Krakow's heritage tourism product and features a walking trail with interpretive boards.

In some parts of CEE, the legacies of Ottoman occupation are slowly being valorized for heritage tourism. The Ottoman period may also be used positively to fashion post-socialist place identities, perhaps the best example being the Stari Most in Mostar, Bosnia (see Grodach 2002). However, although some pasts are being retrieved and included in the heritage product, others remain largely excluded. Examples here include the heritage of the Roma people, which receives relatively little official support, but may continue in less formal and intangible heritage processes such as customs and oral traditions. The government of Macedonia is one of the few examples of the state seeking to preserve Roma heritage through projects focused on music and oral tradition.

Throughout CEE, the primary motive for the promotion of heritage tourism has been economic. These countries have been eager to promote themselves as destinations for heritage tourists from Western Europe. The economic benefits of tourism as a source of income and employment are well established, but they assumed additional significance for the region during a period of major economic restructuring and the dismantling of a centrally planned and state-owned economy. The revenues generated by tourists' spending in general—and heritage tourists in particular—had a significant role to play in improving the balance of payments, in addition to generating currency to fund imports (Hall 2001).

However, while they have been keen to support heritage tourism, CEE governments have not always attached great importance to the preservation and conservation of historic buildings, monuments, and landscapes. In the context of major economic restructuring, post-socialist administrations have faced restricted budgets and numerous conflicting demands on state funds. In these circumstances, state support for the upkeep or restoration of historic buildings has been limited or non-existent, and those resources that are available have often been concentrated on a few showpiece sites (especially those popular with Western tourists) or places of special national significance. At the same time, there is sometimes a lack of an adequate legal framework for protecting historic buildings, towns, and landscapes. Socialist-era legislation was inappropriate, but the drafting of new laws for heritage protection and conservation has not been a priority for many post-socialist administrations. Moreover, the whole notion of the state being involved in heritage conservation was discredited through its association with the centralized planning of state socialism (Hammersley and Westlake 1994). The result is that some historic buildings in the region have continued to suffer neglect and decay through lack of funds for upkeep. There are also entire historic quarters in some cities that have suffered a similar fate. Bucharest's historic center was long neglected during the socialist era—a trend that continued after 1989 because of uncertainties over property ownership and a lack of funds from central and city budgets. Only in 2006 was a comprehensive program for the rehabilitation of the area launched (Chilianu 2006) that could significantly enhance the city's heritage product.

There are other pressures on the historic environment that are specific to the conditions of post-socialism. There are many instances of historic buildings that were nationalized during the socialist era that have subsequently been returned to their former owners or their descendants. In some cases, the new owners lack the finance for maintenance and allow the buildings to decay. In other cases, the new owners, in search of a quick profit, sell the building to private organizations that may not be concerned to maintain its historic character. In addition, the burgeoning free market makes its own demands on the historic environment. Historic buildings—particularly in major urban areas—are in demand as shops, offices, and residences. At the same time, new business premises have been built in many historic cities that are unsympathetic to local architectural traditions (Hammersley and Westlake 1994). Moreover, the visual esthetic of many historic city centers has been substantially changed through the arrival of advertising boards (almost unknown under socialism) and neon signs promoting global brands (Gallagher and Tucker 1996). City planners have had little ability or inclination to restrict such developments.

The heritage of rural areas is under similar pressure. The nature of rural life has changed dramatically since the fall of socialism. Rural areas have experienced massive outmigration of young people, leaving only a fast-aging population maintaining the rural traditions that feature so prominently in promotional materials. Moreover, rural landscapes, particularly those surrounding major cities, face unprecedented (and often unregulated) development pressures. Populations compelled to live in apartment blocks during the socialist era are now rejecting apartment living. Those with the ability to do so—initially the new class of business people, but increasingly the middle classes—are leaving the cities for newly built houses in the countryside. Moreover, many heritage landscapes are now being changed owing to the massive demand among the business class for second and holiday homes.

While economic motives were central to the promotion of heritage tourism in post-socialist CEE, this activity was also underpinned by a significant political and ideological agenda. In particular, heritage tourism had the potential to make an important contribution to post-socialist identity-building. During the four decades of state socialism, the CEE states had developed distinct identities characterized by socialism, an orientation eastwards toward the Soviet Union, and an ideological hostility toward the West. From the perspective of the West, CEE was emphatically Other, characterized by a political and economic system that was the antithesis of the Western model. In fact, this way of viewing CEE as a strange, marginal and, in many ways, less developed periphery is deeply embedded in the Western imagination (Todorova 1997; Wolff 1994). Following the collapse of state socialism, almost all of the CEE states (with a few exceptions, such as Belarus) have embarked on a "return to Europe" with membership of the EU and the North Atlantic Treaty Organization (NATO) being the ultimate aspiration. They have therefore been concerned with creating new identities that are underpinned by the economic and political orthodoxies of Western Europe (Young 2005a). Furthermore, they have sought to project these identities to the West to ameliorate and enhance their images (Morgan and Pritchard 1998) and, in effect, to present themselves as "Same" rather than "Other" (Light 2001).

Heritage tourism has played a key role in this post-socialist reimaging of CEE. In seeking to legitimate their aspirations to EU membership, the CEE states have stressed long-standing historical and cultural ties with the West that were only temporarily severed during the period of state socialism. For example, many countries and cities have highlighted their architectural heritage as a way of demonstrating what they have in common with Western Europe, as there are architectural styles (such as Baroque) that are common to Western, Central, and Eastern Europe. Slovenia has promoted its Habsburg heritage to assert historical ties with Austria and Italy (Hall 2002). Similarly, Romania has stressed its Roman heritage to emphasize historical ties with its Latin "cousins" (France, Italy, and Spain) in Western Europe (Light 2006). In some cases, historic cities in CEE have sought to reposition themselves in the Western imagination by overtly making comparisons to a familiar place in Western Europe. Thus, Romania's capital, Bucharest, is promoted as the "Paris of the East," while Haapsalu, Estonia, is promoted as "the Venice of the Baltics" (Morgan and Pritchard 1998) as is St. Petersburg, Russia. Many post-socialist countries promote a heritage associated with the myth of a pre-socialist, cosmopolitan "golden age" that often recalls an era of greater "connectedness" with Europe. The city of Vilnius (Lithuania) reaches back to the heritage of the Grand Duchy of Lithuania (1300–1500s) in its post-socialist identity construction (Munasinghe 2005); Łódź (Poland) stresses its nineteenth- and early twentieth-century history of economic dynamism and multiple European nationalities (Young and Kaczmarek 1999, 2008); and Banska Bystrica (Slovakia) has focused on the reconstruction of heritage associated with its democratic era of 1918–38 (Bitusikova 1998). In a similar way, although CEE has diverse religious traditions rooted in the Catholic, Orthodox, and Protestant churches, there has been a strong emphasis on promoting this Christian heritage (churches, cathedrals, and monasteries) as another way of asserting shared traditions and values with Western Europe.

In addition to its contribution to post-socialist identity building, heritage tourism has also been an important means for the young and recently created states of CEE to demonstrate and affirm their new independent national identities. Upon attaining independence, each state has sought to establish a new national history as one means to legitimate their existence. In the process, historic buildings have been reinterpreted to become "national heritage." This heritage is then valorized and promoted to foreign tourists (both from Western Europe and also from neighboring CEE countries) as a means to demonstrate both an independent national identity but also membership of, and allegiance to, a common European home. There are many examples in the region. For example, after Slovakia attained independence in 1993, the capital city, Bratislava, underwent an extensive program of restoration and refurbishment to create a showpiece historic city appropriate for a newly independent state. In Croatia, the historic city of Dubrovnik (a World Heritage Site) suffered extensive damage in 1991 during the conflicts that followed the break-up of Yugoslavia. Such was the city's historic significance for Croatia that, once the conflicts had ceased, the damage to the city was almost entirely repaired by 2005 (BBC 2005).

Heritage also plays an important role in promotional activities that go beyond tourism. These include place marketing campaigns at a range of scales to attract foreign direct investment and business relocation (Young 2004, 2005b; Young and Kaczmarek 1999). In particular, through their place promotion, many CEE countries actively draw on pre-socialist heritages that have commonalities with "Western Europe." Examples include the common architectural heritage of the Austro-Hungarian Empire or that of the formerly Hanseatic cities on the Baltic coast. The use of this pre-socialist "European" heritage has two main discursive goals. The first is to cast off associations with the state-socialist era and to reimage these countries as modern, international, capitalist, European, and in tune with the dominant neo-liberal agenda (Young and Kaczmarek 2008). This is important in efforts to dispel the perception of these countries as struggling with the legacies of state-socialist economic systems and societies. Drawing on notions of a "common European heritage" is part of a discursive strategy to signal that the CEE countries are part of Europe and, as such, are firmly embedded in normative notions of the "correct" way to "do business." Second, the use of this heritage is important in "quality of life" place marketing strategies. Portraying cities as essentially European in appearance, character, and values makes them more attractive to transnational professionals, management, and labor who might be persuaded to accompany the movement of key business functions to take advantage of cheaper costs and other factors offering competitive advantage in CEE.

Conclusions

The turbulent and complex past of CEE means that the region has an exceptionally diverse range of heritage resources. These encompass built heritages, including historic towns and cities, a wide range of historic buildings (ranging from the prehistoric period to the socialist era), and commemorative monuments from various historical periods. There are also many sites associated with the region's Christian religious heritage, although in some parts of Southeast Europe, there are also places that reflect a history of Islamic influence. Many rural areas still support long-standing lifestyles and traditions from historical periods that do not have any equivalents in the heritage product of Western Europe. There is also a wide diversity of physical landscapes, ranging from the plains of northern Europe through the Carpathian and Balkan mountain ranges to the Danube Delta. These material heritages are complemented by the many intangible heritages of memories and personal associations with different experiences of the past. Overall, CEE has much to offer for heritage tourism and, since the collapse of socialist regimes, such tourism has increased significantly in the region. This growth was initially generated by Western Europeans but, in recent years, citizens of the CEE states themselves are increasingly participating in heritage and cultural tourism within the region.

While some of these countries are nearly as wealthy as Western European ones, heritage tourism offers an important economic resource to support the development strategies of the CEE countries. Linked to this is the importance of the political and ideological role of heritage tourism in supporting development in the region. Many of these CEE countries are recent creations and are actively engaged in processes of nation- and state-building and socio-economic development in which the selective use of heritage plays an important role. Key elements of their histories have been recovered as "heritage" to create new national myths. Furthermore, the creation of new images and identities for post-socialist places is also important in presenting these newly emerging countries to other states and also to supranational organizations such as the EU and NATO. Again, heritage is mobilized in these processes, particularly the preservation and promotion of what is deemed to be a shared "European" heritage, something that is being used to re-establish the "European" credentials of these newly independent countries. In particular, a key goal of using heritage in these "Europeanization" strategies is the rejection of the socialist past although, paradoxically, the heritage of state socialism is increasingly becoming an attraction for heritage and cultural tourists from the West.

The newly emerging heritages discussed above are, of course, examples of dissonant heritage (Tunbridge and Ashworth 1996). The complex pasts of German, Jewish, Russian, Soviet, and Roma peoples, and of the diverse and complex minority groups found within post-socialist states, make the process of selecting and interpreting heritage problematic. Inevitably, this raises the question of whose heritage is being interpreted, by whom, and for whom? (Tunbridge 1994). There are also clear moral and ethical issues around commodifying traumatic pasts such as concentration camps, socialist prisons, and the actions of state security services (see Ashworth and Hartmann 2005). The socialist heritage that has been produced currently lacks a balanced view of the complex and diverse experience of socialism across different countries at different times, often focusing instead on Western stereotypes involving trauma, repression, and terror. Drawing attention to the heritage of the socialist past may disrupt attempts to draw a line under the socialist period and construct new post-socialist identities. There is also a danger that these countries may continue to be associated in the Western imagination with the unwanted stereotype of "communism." Here, the countries of CEE share some of the issues faced in the development of heritage in other developing countries that are dealing with the legacies of Empire and their experience of colonialism. As such, the planning and management of heritage tourism will continue to present a challenge for the region's planners, politicians, and tourism practitioners, but also for those tourists who are the consumers of this heritage.

References

Ashworth, G. and Hartmann, R. (eds) (2005) *Horror and Human Tragedy Revisited: The Management of Sites of Atrocities for Tourism.* New York: Cognizant Communication Corporation.

Ashworth, G.J. and Tunbridge, J.E. (1999) Old cities, new pasts: heritage planning in selected cities of Central Europe. *GeoJournal*, 49: 105–16.

Bitusikova, A. (1998) Transformations of a city centre in the light of ideologies: the case of Banska Bystrica, Slovakia. *International Journal of Urban and Regional Research*, 22: 614–22.

Boia, L. (1999) *Mitologia ştiinţifică a comunismului.* Bucureşti: Humanitas.

BBC (2005) Adriatic pearl recovers its lustre. *BBC News*, February 1, 2005. Available from http://news.bbc.co.uk/1/hi/world/europe/4223859.stm (accessed August 5, 2007).

Carter, F.W. and Turnock. D. (eds) (1996) *Environmental Problems in Eastern Europe*, 2nd edn. London: Routledge.

Cavalcanti, M.B.U. (1997) Urban reconstruction and autocratic regimes: Ceauşescu's Bucharest in its historic context. *Planning Perspectives*, 12: 71–109.

Chilianu, D. (2006) Incepe refacerea centrului istoric al Capitalei. *Adevărul*, July 21: A8.

Cohen, J.-L. (1995) When Stalin meets Haussmann. The Moscow Plan of 1935. In D. Ades, T. Benton, D. Elliott and I.B. Whyte (eds), *Art and Power. Europe under the dictators 1930–1945*. London: Hayward Gallery.

Czepczyński, M. (2008) *Cultural Landscape of Post-Socialist Cities: Representation of Powers and Needs*. Aldershot: Ashgate.

Dawson, A.H. (1999) From glittering icon to … . *Geographical Journal*, 165: 154–60.

Dingsdale, A. (1999) Redefining "Eastern Europe": A new regional geography of post-socialist Europe. *Geography*, 84: 204–21.

Forest, B. and Johnson, J. (2002) Unraveling the threads of history: Soviet-era monuments and post-Soviet national identity in Moscow. *Annals of the Association of American Geographers*, 92: 524–47.

Forest, B., Johnson, J. and Till, K. (2004) Post-totalitarian national identity: public memory in Germany and Russia. *Social and Cultural Geography*, 5: 357–80.

Frank, S. (2006) Disputed tourist stages, heritage stories and cultural scripts at Checkpoint Charlie, Berlin. In S. Schröeder-Esch and J.H. Ulbricht (eds), *The Politics of Heritage and Regional Development Strategies – Actors, Interests, Conflicts*, pp. 175–85. Weimar: Bauhaus Universität Weimar.

Gallagher, J.J. and Tucker, P.N.J. (1996) Revolution and continuity in the cultural landscape: Transylvania in transition. In D. Turnock (ed.), *Frameworks for Understanding Post-Socialist Processes*, Occasional Paper No. 36, pp. 21–26. Leicester: Leicester University Department of Geography.

Grodach, C. (2002) Reconstituting identity and history in post-war Mostar, Bosnia-Herzegovina. *City*, 6: 61–82.

Hall, D.R. (1991) Evolutionary pattern of tourism development in Eastern Europe and the Soviet Union. In D.R. Hall (ed.), *Tourism and Economic Development in Eastern Europe and the Soviet Union*, pp. 79–115. London: Belhaven.

—— (1995) Tourism change in Central and Eastern Europe. In A. Montanari and A. M. Williams (eds), *European Tourism: Regions, Spaces and Restructuring*, pp. 221–44. Chichester: Wiley.

—— (1998) Tourism development and sustainability issues in Central and South-eastern Europe. *Tourism Management*, 19: 423–31.

—— (2001) Tourism and development in communist and post-communist societies. In D. Harrison (ed.), *Tourism and the Less Developed World: Issues and Case Studies*, pp. 91–107. Wallingford: CABI.

—— (2002) Brand development, tourism and national identity: the re-imaging of former Yugoslavia. *Brand Management*, 9: 323–34.

Hall, D. (2004) Rural tourism development in Southeastern Europe: transition and the search for sustainability. *International Journal of Tourism Research*, 6: 165–76.

Hammersley, R. and Westlake, T. (1994) Urban heritage in the Czech Republic. In G. J. Ashworth and P.J. Larkham (eds), *Building a New Heritage: Tourism, Culture and Identity in the New Europe*, pp. 178–200. London: Routledge.

Henderson, J.C. (2007) Communism, heritage and tourism in East Asia. *International Journal of Heritage Studies*, 13: 240–54.

Herrschel, T. (2006) *Global Geographies of Post-Socialist Transition: Geographies, Societies, Policies*. London: Routledge.

Kim, S.S., Timothy, D.J. and Han, H.-C. (2007) Tourism and political ideologies: a case of tourism in North Korea. *Tourism Management*, 28: 1031–43.

Koonz, C. (1994) Between memory and oblivion: concentration camps in German memory. In J.R. Gillis (ed.), *Commemorations: The Politics of National Identity*, pp. 258–80. Princeton, NJ: Princeton University Press.

Jansen, O. (2005) Trauma revisited: the Holocaust Memorial in Berlin. In G. Ashworth and R. Hartmann (eds), *Horror and Human Tragedy Revisited: The Management of Sites of Atrocities for Tourism*, pp. 163–79. New York: Cognizant Communication Corporation.

Light, D. (2000a) Gazing on Communism: heritage tourism and post-communist identities in Germany, Hungary and Romania. *Tourism Geographies*, 2: 157–76.

—— (2000b) An unwanted past: contemporary tourism and the heritage of communism in Romania. *International Journal of Heritage Studies*, 6: 145–60.

—— (2001) Facing the future: tourism and identity-building in post-socialist Romania. *Political Geography*, 20: 1053–74.

—— (2006) Romania: national identity, tourism promotion and European integration. In D. Hall, M. Smith and B. Marciszewska (eds), *Tourism in the New Europe: The Challenges and Opportunities of EU Enlargement*, pp. 256–69. Wallingford: CABI.

Light, D. and Dumbrăveanu-Andone, D. (1997) Heritage and national identity: exploring the relationship in Romania. *International Journal of Heritage Studies*, 3 (1): 28–43.

Lubonja, F. (2002) *Nostalgia. Essays on the Nostalgia for Communism.* Sękowa, Poland: Czarne Publishing.

Morgan, N. and Pritchard, A. (1998) *Tourism Promotion and Power: Creating Images, Creating Identities.* Chichester: Wiley.

Munasinghe, H. (2005) The politics of the past: constructing a national identity through heritage conservation. *International Journal of Heritage Studies*, 11: 251–60.

Murzyn, M. (2006) *Kazimierz. Środkowoeuropejskie doświadczenie rewitalizacji* (*The Central European Experience of Urban Regeneration*). Kraków: International Cultural Centre Kraków.

Nawratek, K. (2005) *Ideologie w przestrzeni. Próby demistyfikacji.* Kraków: Universitas.

Olszański, T. (1970) *Budapeszteńskie ABC.* Warszawa: Iskry.

Prahoveanu, I. and Coşuleţ, S. (1985) *Muzeul Bran, Ghid.* Braşov, Poland: Comitetul de Cultură şi Educaţie Socialistă al Judeţului Braşov.

Rembowska, K. (1989) Przestrzeń znacząca. *Miasto Socjalizmu Kwartalnik Geograficzny*, 4: 11–14.

Satjukow, S. and Gries, R. (eds) (2002) *Soziliastische Helden. Eine Kulturgeschichte von Propogandafiguren in Osteuropa und der DDR.* Berlin: Ch. Links.

Sidorov, D. (2000) National monumentalization and the politics of scale: the resurrections of the Cathedral of Christ the Savior in Moscow. *Annals of the Association of American Geographers*, 90: 548–72.

Stenning, A. (2005) Post-socialism and the changing geographies of the everyday in Poland. *Transactions of the Institute of British Geographers*, 30: 113–27.

Timothy, D. (2001) *Tourism and Political Boundaries.* London: Routledge.

Todorova, M. (1997) *Imagining the Balkans.* Oxford: Oxford University Press.

Tunbridge, J.E. (1994) Whose heritage? Global problem, European nightmare. In G.J. Ashworth and P.J. Larkham (eds), *Building a New Heritage: Tourism, Culture and Identity in the New Europe*, pp. 123–34. London: Routledge.

Tunbridge, J.E. and Ashworth, G.J. (1996) *Dissonant Heritage: the Management of the Past as a Resource in Conflict*, Chichester: Wiley.

UNDP (United Nations Development Programme) (2006) *Human Development Report 2006.* New York: United Nations Development Programme. Available from http://hdr.undp.org/en/media/hdr06-complete.pdf (accessed January 13, 2008).

UNESCO (2009) World Heritage List. Available from http://whc.unesco.org/en/list (accessed February 2, 2009).

Wolff, L. (1994) *Inventing Eastern Europe: The Map of Civilisation on the Mind of the Enlightenment.* Stanford, CA: Stanford University Press.

World Bank (2008) Available from http://siteresources.worldbank.org/DATASTATIST ICS/Resources/CLASS.XLS (accessed 7 February 2008).

World Tourism Organization (2006) *Tourism Market Trends, 2006 Edition.* Available from http://www.unwto.org/facts/eng/pdf/indicators/ITA_europe.pdf (accessed January 19, 2008).

Young, C. (2004) From place promotion to sophisticated place marketing under post-socialism: the case of *CzechInvest. European Spatial Research and Policy*, 11: 71–84.

—— (2005a) Post-socialist East and Central Europe. In M. Phillips (ed.), *Contested Worlds*, pp. 251–86. Aldershot: Ashgate.

—— (2005b) Meeting the new foreign direct investment challenge in East and Central Europe: place-marketing strategies in Hungary. *Environment and Planning C*, 23: 733–57.

Young, C. and Kaczmarek, S. (1999) Changing the perception of the post-Socialist city: place promotion and imagery in Łódź, Poland. *Geographical Journal*, 165: 183–91.

—— (2008) The socialist past and post-socialist urban identity in Central and Eastern Europe: the case of Łódź, Poland. *European Urban and Regional Studies*, 15: 53–70.

Young, C. and Light, D. (2006) "Communist heritage tourism": between economic development and European integration. In D. Hassenpflug, B. Kolbmüller and S. Schröeder-Esch (eds), *Heritage and Media in Europe – Contributing towards Integration and Regional Development*, pp. 249–62. Weimar: Bauhaus Universität Weimar.

14 Heritage tourism in the developing world

Reflections and ramifications

Although heritage in the developing world is no less important than in the developed world, the language of heritage used today reflects a Western bias (Lowenthal 1997). Heritage tourism is a global phenomenon and has been in practice for thousands of years in the form of pilgrimage, but the tourism literature is dominated by encounters between Westerners and their hosts in developing countries (Winter 2007).

Despite the abundance and uniqueness of heritage resources in the developing world, the heritage tourism framework used in less-developed countries (LDCs) is influenced by Western-centric models (Winter 2007). Even UNESCO's World Heritage program is designed essentially from a Western perspective. For example, the concept of World Heritage started in Europe in 1931, followed by Tunisia, Mexico, and Peru in 1964, and finally eighty nations from all continents joined the World Heritage Convention in 1979 (Lowenthal 1997). Most of the literature on heritage in developing countries is written by Westerners, who are often biased and misrepresentative. Western authors often describe the "East" and "South" as primitive and remote, and any change that occurs is considered negative. While biases are always present in academic works, this book has attempted, inasmuch as possible, to involve authors from the developing regions being considered. Despite earnest efforts, some of our arrangements with local authors fell through. Nonetheless, despite the limited number of scholars working on heritage and tourism in developing countries, we were successful in putting together a team of authors who represent well the nine major world realms being represented in this tome.

With a collection of nine chapters from various regions of the developing world and an additional four overview chapters, the aim of this book is to examine the existing paradigms and issues that developing countries are facing in the realms of heritage and heritage tourism. Because each developing country has its own history, culture, geography, religions, politics, economy, and of course heritage and tourism, it was unrealistic to discuss each country; for this reason, global realms were identified and issues written that are common within each realm or region.

Compared with the size and population of developing countries, their share of international tourist arrivals and receipts are low. The share of tourist arrivals to developing regions in relation to global arrivals is as follows: South Asia 1.6 percent, South East Asia 5.5 percent, the Pacific 3.6 percent, Central and Eastern Europe 5.1 percent, the Caribbean 3.0 percent, subSaharan Africa 2.2 percent, North Africa and Middle East 5.9 percent, and Latin America 2.6 percent (World Tourism Organization 2007). Travel to and from Europe and North America, on the other hand, accounts for approximately 79 percent of all international trips and about 78 percent of all receipts (World Tourism Organization 1999). The reasons for such a disparity in travel between developed and developing regions are related to political instability and conflict, health concerns, poor facilities and infrastructure, inadequate levels of service, inadequate distribution channels, lack of knowledge in potential markets, and the high cost of travel to many of the world's more peripheral places (Gartner and Lime 2000).

Compared with other forms of tourism, heritage tourism in the developing world has a more intraregional and domestic focus, and the trend toward tourists from LDCs traveling to other LDCs is growing. However, tourism statistics compiled by the UNWTO do not take into account the size of this market for two main reasons. First, many countries have cultural and political ties and do not require visas for people crossing common borders, such as India and Nepal. Many Indian tourists travel to Nepal to visit shrines, temples, and religious events. However, because they do not require a visa, or even an identity card, there are no records of Indians visiting Nepal by land. Second, domestic and regional tourists have different values and behaviors than Western tourists. The tourism industry is more focused on Western tourists and their needs, and fails to recognize the economic contributions of domestic tourists. Although per capita spending by domestic and regional pilgrims is significantly lower than that of other tourists, their overall economic contribution should not be ignored.

The regional and domestic heritage and pilgrimage market also helps diversify the tourism product and reduce the impacts of low seasonality. Thus, there is a need for regional cooperation and collaboration in promoting and developing heritage tourism, not only for widespread global audiences but for regional markets as well (Timothy 2003).

The need for regional cooperation in each part of the developing world has been felt since these countries' independence in the 1950s, 1960s, and 1970s. As a result, nearly all countries have joined various supranational alliances, such as ASEAN (the Association of Southeast Asian Nations), SADC (Southern African Development Community), SAARC (South Asian Association for Regional Cooperation), ECOWAS (the Economic Community of West African States), and CARICOM (Caribbean Community) in an effort to widen their global competitive advantage. Although there have been some tourism initiatives associated with these associations, tourism has not been a major focus (Ghimire 2001), as it has been in the developed world context (e.g., the North American Free Trade Area (NAFTA) and the EU). With their concern about cross-border travel, environmental conservation, economic development, and intraregional transportation, these organizations have the potential to play an important role in promoting regions at large, facilitating the movement of tourists, and considering the value of regional heritage products (Timothy 2003).

This book identifies poverty as a major issue affecting the preservation of heritage and the development of heritage tourism in underdeveloped nations. Unfortunately, but understandably, heritage conservation is afforded a low priority in countries and regions where the majority of the population struggles to survive. This is especially evident in Africa, Latin America, and parts of Asia, where low per capita incomes and other development indices tend to correspond with low levels of tourism and a lack of conservation efforts.

Budget problems plague most of the world, but these are especially acute in less-affluent regions, where money is rarely allocated for heritage preservation; however, the issue is not just a matter of budgets, but also a matter of public agency priorities. In many cases, such as in India, there is simply too much patrimony to conserve, given meager budgets and human resources. All too often, governments focus more on building new mega-structures in lieu of preserving older, more traditional structures. For example, in Bagan, Myanmar, new pagodas have been built atop ancient ruins using inauthentic materials in the name of restoration (see Chapter 5). A giant Buddha statue is currently under construction in Thimpu, Bhutan, while many historic buildings are falling apart. The Indonesian government has built a giant *Garuda Vishnu Kencana* statue as a tourist attraction in Bali, while many communities fight to save their traditional villages.

Corruption compounds these monetary scarcities even further. Corruption is prevalent in the developing world and takes place in different forms. Bribery is the most common form, affecting the daily lives of ordinary citizens and dictating how the public sector operates. According to the most recent report by Transparency International (2008), less-developed countries have the highest corruption perceptions index (CPI). The report also shows a strong correlation between poverty and corruption. Corruption overshadows more than financial issues, as it takes place in other forms such as political corruption, abuse of power, and favoritism in many sectors including health, education, and trade in illicit antiquities. Further, the efforts of poor countries to alleviate poverty are often plagued by dishonest judiciaries, political parties, and bureaucracies. Unfortunately, more than half of the citizens surveyed by Transparency International around the world expect the level of corruption to increase in the future. Bribery and other forms of corruption influence what heritage products are selected for show, financed for conservation, and traded on the world market.

There is also a general lack of understanding and focus regarding intangible heritage. Intangible heritage is eroded by globalization and modernization processes. There is a major ongoing debate about the costs and benefits of the preservation of intangible heritage versus modernization and, unfortunately in LDCs, the two concepts are usually seen as mutually exclusive and

incompatible. While many, mostly Westerners, view change and modernization as primary causes of lost "primitiveness", many native people see it as a
sign of progress, which enhances their socio-economic condition and quality
of life (see Chapter 5).

As discussed in several chapters in this book, developing countries are very
rich in heritage; however, the linkage between heritage and tourism is weak.
One of the potential ways to strengthen the linkage is interpretation, as Nuryanti (1996: 251) described, "the less developed countries – in the worlds of
traditions, cultures, religions, superstitions and distance from modernity –
have the potential to be rediscovered as a source of symbols and new interpretations." One of the fundamental challenges the developing countries face
is that heritage portrayed by Westerners is unreal and stereotypical (see
Chapter 7). Heritage is often falsified in the production and (mis)representa-
tion of culture (MacCannell 1992). This issue can only be resolved through
indigenous interpretation, not interpretation by outsiders. Interpretation is
more than a description of facts; it should include context-specific truths,
indigenous voices, emotional responses, deeper meanings and understandings,
and ownership of the people who own the heritage (Nuryanti 1996). The
outcomes of interpretation should also create greater appreciation, awareness,
understanding, self-fulfillment, and enjoyment for visitors (Herbert 1989).

Other issues common to many developing countries are political instability
and conflict. Remnants of the past observed today survived through historical
violence and conflicts. It is uncertain how much has already been lost because
of contestation and violence, but it certainly must be immense. Minority and
indigenous heritage is especially vulnerable. For example, the indigenous
heritage of the Caribbean was essentially wiped out during the violent conquest of the islands in the sixteenth century and the slavery that followed (see
Chapter 11). Even today, many Buddhist sites in South Asia are located in
areas inhabited by other religious majorities and face potential annihilation as
religious and racial relations continue to deteriorate (see Chapter 8).

Colonialism is another common characteristic of most developing countries. It has affected the current states and their heritage in many ways, not
least of which is political instability as administrative regions were carved out
by outsiders with little regard for socio-cultural and religious boundaries (see
Chapters 9, 10, and 12). In addition, it was not uncommon for some colonial
powers to adopt assimilationist policies in their colonies that dictated the
demise of indigenous culture and religion in favor of the cultures, politics, and
religions from Europe. Thus, in some places, little remains of the original
heritage, but colonial heritage abounds. Patrimonial contention also exists in
places where forced migration occurred, as in the case of slavery, which has
created heritage identity crises for entire tribes, races, and nations.

Despite these challenges, heritage managers and governments have a vast
menu of opportunities for using heritage tourism to enhance the quality of life
of destination residents. Unfortunately, a pattern exists in less-developed
regions where large numbers of tourists arrive but the benefits of their

spending are not transmitted to the site and the surrounding community; instead, these communities and historic locations are relegated to bear the burdens of mass tourism. Heritage, above many other resources, should become a brighter beacon in alleviating poverty through tourism development, but all too often it is ignored or misused.

A primary reason for this is that some heritage managers see tourism as a problem rather than a tool for finding solutions. There is an unfortunate, albeit widespread, sense of skepticism among heritage managers regarding tourism. There is a concomitant lack of understanding among heritage managers that proper planning and appropriate tourism-related uses of the past can help minimize the negative effects of tourism and maximize its benefits for the community and for their conservation efforts. Heritage managers must begin thinking about the long-term viability of the past rather than simply seeking short-term assistance from international aid agencies. Although major projects such as land acquisition are costly and need large-scale investments from governments and aid agencies, most of the time, heritage is threatened simply by a lack of even minimal budgets to maintain and manage.

This book clearly demonstrates that the challenges developing countries face in the realm of cultural heritage tourism are different and, in some cases, more intense than problems in more affluent countries. In the Western world, heritage concerns relate more basically to normal planning and management, including building demolition, traffic management, and parking problems, whereas in the poorer countries, the issues are more complicated, involving very profound problems related to forced relocations, inadequate compensation, and lack of resources and institutional capabilities (Nuryanti 1996; Nyaupane 2009).

Most heritages are created through long historical processes, but there are some instances, such as in Eastern Europe and South Asia, where heritages are created quickly and are constantly evolving as one set of ideologies is replaced by another. Heritage and heritage tourism are extremely complex phenomena, particularly in the less-developed world. This situation provides tourism scholars, policy makers, and historic site managers with an open laboratory in which to study the evolution of heritage, places, and meanings.

One of our goals in putting together this collection was to raise more questions than we answer. This has been a major success. There are many questions still unanswered. Who owns the heritage and who does not? What implications does this have for conflict resolution? How can heritage be used as a resource to advance pro-poor tourism more effectively? How can the heritage of today be utilized in such a way that it will still be around for future generations? Although some of this work appears to take a negative stance, examining challenges more than opportunities, this is not by choice. It is simply reflective of the concepts and issues researchers have identified in the less-affluent parts of the world. Nonetheless, there is clearly considerable scope for research on improving heritage conditions, overcoming political obstacles, understanding new trends in heritage tourism, appreciating domestic forms of heritage-based travel, the richness and depth of various cultures, and how tourism functions within different contexts, and appreciating new opportunities that come to light.

There is a clear need for more comprehensive research in understanding heritage tourism issues in the less-developed world. This collection of essays only begins to scratch the surface. It is our hope that students and researchers of tourism will join the effort to understand and enhance the role of heritage as a resource for sustainable tourism in the developing parts of our world.

References

Gartner, W.C. and Lime, D.W. (2000) The big picture: a synopsis of contribution. In W.C. Gartner and D.W. Lime (eds), *Trends in Outdoor Recreation, Leisure and Tourism*, pp. 1–13. New York: CABI.

Ghimire, K.B. (2001) Regional tourism and south–south economic cooperation. *The Geographical Journal*, 167(2): 99–110.

Herbert, D.T. (1989) Leisure trends and the heritage market. In D.T. Herbert, R.C. Prentice and C.J. Thomas (eds), *Heritage Sites: Strategies for Marketing and Development*, pp. 1–15. Aldershot: Avebury.

Lowenthal, D. (1997) *The Heritage Crusade and the Spoils of History*. New York: Viking.

MacCannell, D. (1992) *Empty Meeting Grounds*. London: Routledge.

Nuryanti, W. (1996) Heritage and postmodern tourism. *Annals of Tourism Research*, 23: 249–60.

Nyaupane, G.P. (2009) Heritage complexity and tourism: the case of Lumbini, Nepal. *Journal of Heritage Tourism*.

Timothy, D.J. (2003) Supranationalist alliances and tourism: insights from ASEAN and SAARC. *Current Issues in Tourism*, 6(3): 250–66.

Transparency International (2008) The 2008 results. Available from http://www.transparency.org (accessed on October 5, 2008).

Winter, T. (2007) Rethinking tourism in Asia. *Annals of Tourism Research*, 34: 27–44.

World Tourism Organization (1999) *Tourism Highlights 1999*. Madrid: UNWTO.

—— (2007) *Tourism Highlights, 2007*. Madrid: UNWTO.

Index